Global Environmental Concerns

Dedicated to My Parents

Shri R. K. Nilamani Sana

And

Smt. Achoubi Shija

Global Environmental Concerns

(With Special Reference to Manipur)

R. K. Ranjan Singh

RAJESH PUBLICATIONS
NEW DELHI - 110 002

First Published in India : 2000
ISBN : 81-85891-32-X

Published by :
M.L. Gupta
RAJESH PUBLICATIONS
1, Ansari Road, Darya Ganj,
New Delhi - 110 002

Type Setter :
Anita Gupta
SANJAY ENTERPRISES
New Delhi-110 002

Printed at :
GRAPHIC PRINT
New Delhi-110 002.

PREFACE

This book *GLOBAL ENVIRONMENTAL CONCERNS (With Special Reference to Manipur)* is a compilation of articles already published and presented in different journals, papers and seminars and workshops during the past 25 years. These papers basically aim at creating awareness among the common masses and specifically among the youths and students. Further, in most of the chapters, special emphasis is being given to the relationship between man and nature, nature and the present day concept of development i.e. 'mastering nature' which is what ecosystem destruction is currently doing, but about cohabiting in harmony with it — conservation of wetlands and biodiversities etc.

What is nature ? What is humanity's place in nature ? And what is the relationship the society has in the natural world ?

In an era of ecological breakdown, answering these questions has become of momentous importance for our everyday live and for the future that we and other life forms are to face. At a first glance, all of us 'know' what nature is and what it means. It is that which are all around us — the flora and fauna, the rocks and minerals and the like. It is those resources which 'we human beings 'are' exploring and exploiting' relentlessly for its own end. Defining 'nature becomes an even more complex task when we include the human species as part of it'. It is also a fact that it is the Nature not Man that provides existence to lives on Earth.

Here, it may be quoted what Chief Seattles wrote in his letter to the American President, Franklin Pierce in 1854 : *"Every part of this earth is sacred to my people. Every shining pine needle, every sandy shore, every mist in the dark woods, every clearing and humming insect is holy in the memory and experience of my people. The sap which courses through the trees, carries the memories of the Red Man.................... The rivers are our brothers, they quench our thrist. The rivers carry our canoes and feed our children.......... The air is precious to the Red Man, for all things share the same breath. Teach your children what we have taught our children that the earth is our mother...... The earth does not belong to man; man belongs to the earth."*

Our thoughts and aspirations being influenced and stimulated by Chief Seattles' mental feelings, we are made to turn our attention to the panicky situation prevailing in the state of Manipur which is having diversed topographical features where the gift of nature is put at a higher scale, both in the hills and plains, as compared to others. Although ours is a small state in the North-Eastern corner of India, we come to realise the fact that nature has given a very rich plant and animal genetic resources in this part of the earth. This state has been the Vavilovian centres of diversity and origin of a variety of Angiospermic, plants yielding food, fodder, fuel, fibre and medicine. There is considerable inflow and outflow of flora and fauna between the Southeast Asian countries and the Indian Sub-continent through this part of the earth. Manipur, incidentally, also happens to be one of the two biodiversity 'hot spots' of the country. The two megacentres of endemism in India are the Western Ghat and the Eastern Himalayas with 26 micro-endemic centres and Manipur occupies a unique position as a number of rare indigenous species are available. Our cultural heritages were fully influenced by our *Natural Heritage*. Our hills, our rivers, and lakes are sacred to all the Manipuris. To preserve and conserve our *Natural Heritages* will be the only means, which could kindle the light of *national identity* of this part of the erath. I hope this book of plain thinking may make a campaign for protection of Environment.

R. K. Ranjan Singh

ACKNOWLEDGEMENTS

I would like to express my grateful thanks to the people of Manipur who allow me to share my views to all issues related with the Environmental concerns of this part of the country. Secondly, dedicated voluntary organizations namely All Manipur Ithai Barrage Affected Peoples' Organization; Global Science Club, Khoijuman; Generation De New Image, Ningthoukhong; Environment Social Reformation and Sangai Protection Forum, Keibul Lamjao; Manipur Association for Science and Society (MASS), Imphal; District Organizaing Committees, Bharat Jan Vigyan Jatha : Thoubal, Bishnupur, Churachandpur, Tamenglong, Ukhrul, Senapati, Chandel and Imphal for providing me a wide varieties of platform of the people participation programme on the changing Environmental concerns in the State. I am also thankful to the NSS Units of the Colleges of the State for providing me to express my views on the issues of Watershed Management and Wasteland Development in the State with all the camper volunteers. It is also thankful to the All India Radio, Imphal for providing me scopes for deliberations on the issues related to the Environmental problems of the State on many occasions. I am also extremely thankful to all concerns that organized these programmes during Second Saturdays, Sundays and holidays so that normal functioning and duties of the Manipur University where I am working was not affected. I am especially thankful to Shri Hanjabam Priyokumar Sharma for his valuable editing and proof reading of this book, and Shri M. Tomba Singh, Stenographer, Manipur University for taking pain of typing the manuscript of the book.

I recorded my deep sense of gratitude to my wife, Smt. Ch. Debala Devi for constant inspiring assistance in many ways and for shouldering herself most of my domestic duties all the time so that I could complete the work.

R. K. Ranjan Singh

Contents

1

Indigenous People of Manipur : Heritage and Biodiversity

Introduction

Through a long history of evolution, the human race has evolved on our Earth in the womb of MOTHER NATURE. It is believed widely that human beings possess the highest intelligence and rationality among all living beings. In the beginning, human behaviour was, by and large, governed by inner instincts and the forces of nature in the struggle for survival. In a way, both inner and outer 'nature' determined all aspects of human life, including its evolution. Inevitably, ancient people lived a life that was inextricably linked with natural objects and phenomena. It was human beings' close relationship with nature, both in terms and struggle against the vagaries of nature and peaceful co-existence with nature, that has led, through a very long period of evolution, to a period of human culture and civilization. Abiding by the laws of nature has been the guiding principle in our march to civilization.

Though we inhabit in a completely different world today, dominated by science, technology, industry and the market, we are still confounded by a number of natural phenomena that remain mysterious. These phenomena show the continuity of human cultural tradition. Some examples of these remnants of this cultural tradition are the worship of stone boulders, caves, trees, lakes, rivers, fire, snakes (reptiles) and cosmic bodies - all of which are held sacred by indigenous people throughout the world. Such reverence for natural elements and objects underpins a fundamental philosophical idea that links our past to our future, in terms of natural, cultural, religious and ethnic heritage. It has bound the human race with nature in a symbolic and mutually sustaining relationship.

Present Development Paradigm

Unfortunately, over the recent couple of centuries, dilemmas, controversies and conflicts have confounded the whole scenario with the emergence of an era of uncontrolled industrial production and consumption. Human consumerism has come in direct conflict with our relationship with nature and its laws.

In fact, most of these conflicts are results of modern national governmental actions and policies that further material production but are detrimental to the well being of organisms, species populations and life communities in the Earth's natural ecological system. To put it another way, such conflicts occur whenever the preservation and protection of natural and wild living beings involve some cost in terms of human benefits. Thus, given the rise of advanced technologies, and an economy dependent on and geared for higher and higher levels of consumption, what is left of the natural environment is quickly disappearing. The more we consume for ourselves, the less there is left for other species. It is a state of clear-cut value conflict - a development model that comes in direct conflict with nature.

We need to view this conflict from an ethical standpoint - as something that exemplifies a moral order. By imposing constraints on our own life style and cultural practices, we, as moral agents, should have the capacity to replace the chaos of a world torn to pieces by human greed. There is no reason why the human race and a great variety of animal and plant life cannot exist side by side on our planet home. But to do so, to regain the "lost paradise", we need to regain the world moral order that allows us to accept the Earth and other species as ONE. We must impose once again, natural bounds for our population expansion, our consumption patterns, and our technological "marvels". If we respect our own lives and our own rights, we must also genuinely respect the lives and rights of all living organisms in the world.

Indigenous Knowledge Systems in Manipur

It is a matter of pride for us that till recently this moral order, a mutually respecting and mutually sustaining relationship with nature, existed among our indigenous people in the tiny state of Manipur. Physiographically, ours is a hilly terrain, with a little piece of alluvial valley dotted with wetlands in the centre.

Manipur has an altitude ranging from 35 mt. a.m.s.l. to 3000 mt. a.m.s.l.; hence, it is blessed by different climatic conditions enriching its bountiful biological diversity. Our small state is famous for its rich heritage of flora and fauna. This, in fact, reflects the highly resourceful gene pool of the region. Manipur has been recognized today as the meeting ground for Indo-Tibetan and South East Asian animal kingdoms, being a juxtaposed zone of the Indian sub-continent and the Indo-China sub-continent.

However, the resourceful natural bounty of this area has been tremendously threatened by the ever-increasing onslaught of market forces blowing across the globe, the complete negligence of indigenous knowledge systems (IKS) and the so-called modernization drives in the area.

For example, the indigenous practice of *paam* (jhum or swidden) cultivation on the hill slopes by both highlanders and lowlanders of Manipur was a good and scientific system of land-use. This was in vogue when the indigenous population was living in tune with the laws of nature. Now-a-days, both the growth of population and the impact of urban-based Western cultural assimilation processes in the area have resulted in great distortions in this perfectly appropriate farming science. Every village community in the region followed an ethic code of conduct for using community lands, whether jointly or individually. The people were compelled to strictly follow certain customary rules like which areas will be under *paam* cultivation for the current season, what types of trees of what age can be cut and which trees will be allowed to grow, etc. There were even areas near the villages where the forest was preserved for other common purposes, and felling strictly of trees regulated.

After the harvest, individual households had to preserve seeds for the next cultivation season. There was no question of a gene bank or a patent rights, and no dependence on the so-called modern seed technology. Every elder woman knew better how to preserve the seed through indigenous treatments of different crop seeds for posterity. Being largely forest dwellers, the people were totally dependent on the forest resources, and there was plenty for survival and leisure.

Together with these occupational practices, many ethical and cultural practices also evolved in the midst of forest ecologies, where all kinds of non–human species lived their natural lives. In the later stages, these practitioners of swidden cultivation formed their own traditional codes of conduct not only for land–use but also for cropping patterns and the preservation of flora and fauna in the areas surrounding their villages. Everything related to our lives, including material needs, health and healing needs, fertility of the people or the earth; and everything else was based on the well being of natural objects in their natural forms. Here, we may assume that both the human and the non–human were all totally and commonly bound by the laws of nature.

Destructive Transitions

Unfortunately, such kind of mutual linking with the nature has been destroyed by the so–called introduction of modern technology and its dubious benefits. This is a direct fall–out of our exogamous education system, whether secular or not. The impact of modern education and religions eroding the traditional value system of the indigenous people in this region has been on the increase particularly in the recent decades. Some kind of critical point seems to have been reached beyond which the winds of destructive trends blow unhindered.

Infact, so enamoured are we by the new–fangled system that most people in our new educational system are not even aware of the indigenous value system and practices. Consequently, most of the modern younger generation disown the gift of nature and biological diversity, the priceless heritage of their ancestors. Today, for our educated elite there is no link between modern education and modern means of life styles and nature. It is a mystery how all is not lost.

Development imperatives within the national framework are also dictated by this education system and its perspective. Large development projects and programmes are politically promoted and infused into our lands. For many such ventures, it is now a question of *quid pro quo*. We have to live with them, the logic goes, as so much of our national resources have been invested the benefits are to a large extent addictive. The much of development has to be hastened and changes must be immediate, determined by five–year planning cycles, otherwise we will be all left behind by others. The thought of what are we leaving behind for our future is never thoroughly understood nor discussed.

Rearguard Actions and Indigenous Conservation Practices

Indigenous people in our region, living in some of the most remote and inaccessible villages are engaged in conserving our natural heritage and its diversity even today. For example, in some of our villages, the people were found to be using at least fifteen (15) different varieties of local hill rice (paddy species) for cultivation. These varieties of paddy were most sustainable for the prevailing local conditions, economic activities and consumption patterns. This traditional, age old methods of cultivation and seed preservation are much more superior and sophisticated than the modern techniques and inputs. The local seeds and plants are free from diseases and pesticides and also from chemical fertilizers. This is one of the most successful stories of the local people and their heritage.

The Manipur Association for Science and Society (MASS), a completely non–funded voluntary organization in Manipur, has attempted to document the biological diversity of the area for future analysis and security. As a result of this documentation attempt, MASS has so far recorded more than two hundred (200) varieties of local rice cultivars with the help of some agricultural scientists. It is presumed that this is one of the richest biodiversity heritage of the indigenous people of Manipur. Apart from the rice cultivars recorded, a large variety of medicinal herbs and plants are still preserved by the local people for their home remedies and life aids.

Similarly, almost all the lowland dwellers in the valley maintain and protect some areas as sacred groves for the village or clan ancestral deity (called *umang lai*). This tradition has been practiced continuously since the evolution of valley settlements, villages and larger habitat agglomerates. It may be mentioned here that the lowlanders (to be distinguished from plains dwellers from continental India), mostly belonging the Meitei people, evolved out of a convergence of seven major and a few other minor clans in historical times. Each clan had its own ancestral deity or deities (actually fore–parents), sacred grove sites for these deities as well as particular totem animals and plants that are considered to be holy and hence, protected religiously.

In their sacred groves or protected forests, no one was permitted to touch the plant or tree or being living in them. This restriction is enforced as a social taboo as well, so the protective mechanisms are quite elaborate and complex. As a result, in the recent past, almost all the forests of the *umang lai* were found to be the ideal zones for the protection and conservation of our priceless biologically diverse heritage. Simultaneously, a number of occupational and cultural activities, or, in short, all kinds of cultural waves and patterns, were generated as a living heritage of the indigenous lowland dwellers of Manipur. All these cultural patterns and ethos clearly reflect an adaptive existence that identifies humans with their natural environment as a single entity. Through these cultural motifs, indigenous people in our region practice meaningful preservation and conservation of the biological diversity which they formed a part of.

Another important practice of the Meitei people which symbolizes and exemplifies the deep concerns for future generations is the ceremony called *ngamu ushin sinba* (the sending forth ceremony of the Meitei *ngamu - chana Slrientalis (schneilder)*– a small indigenous fish). In this ceremony of ancient origin, fingerling sized *ngamu* with two corns of rice (one red and one white) in their mouth are released into the water of rivers, ponds or other wetlands. The ceremony is performed during the month of *Lamda* (March–April) on a Saturday called *Lamtai Thangja*. By this sending forth ceremony, every individual member of the family hoped that they will be protected from all unfortunate circumstances during the ensuing year. This is yet another success story in the traditional conservation of biodiversity within an indigenous knowledge system, embedded as it is within a household cultural practice. The *ngamu* is a very important fish, its flesh known to have great healing and invigorating powers, in the diet of the Meitei people. The fish is used during recuperation from illness, for the weaning of infants, for the elderly.

Among our indigenous people, there are important ceremonies and rituals associated with the felling of trees for various purposes. Among the *Imemi* (Mao people), there is an elaborate ritual that lasts several weeks before the felling of a great tree which is to be used to make the gate to the village – the *koro.* Numerous suchlike rituals are in existence among our people, before cutting bamboo, plantain, etc., and even before the plucking the leaves of many plants. This shows an intimate relationship with the growing trees and plants. Many plants and trees are planted within the homestead lands by cultural norms and practices too. The people felt proud of their plants and flowers.

Emerging International Focus and Mechanism

International understanding about these issues is growing with the adoption of the Convention on Biological Diversity by the United Nations. The contracting Parties (national governments) to this Convention, during its third meeting in Buenos Aires, Argentina in 1996, discussed the important but complicated issues surrounding the spirit, intent and measures to be taken to implement the provisions relating to indigenous and local communities provided in article 8(j). The issues that are focused on article 8(j) and related articles are contentious because of linkages with technology transfer, access to genetic resources, ownership, indigenous peoples' rights, alternative systems of protection of knowledge, innovations and practices of indigenous and local communities embodying traditional life styles, incentives and many other issues that relate to world market forces. The concept of sustainable use of biological diversity is one that has been interpreted in various ways, very few of which are true to the ancient indigenous ethos and practices.

Furthermore, the International Labour Organization's (ILO) Convention No. 169 (Indigenous and Tribal Peoples, Convention 1989) also clearly recognizes the rights of indigenous people relating to land, in Articles 13 to 19.

It is again rather unfortunate that governments are taking sole rights to decide on our future and the heritage of our generation to come. The call from the indigenous people's movement in the United Nations to critically examine all the ramifications of the Convention on Biological Diversity and its implementation through the establishment of mechanisms that facilitate the participation and recognize the rights of indigenous people worldwide needs to be supported. The Buenos Aires conference adopted this suggestion in its decision (III/4), but the process of indigenous people's participation still needs to be strengthened.

Conclusion

Living practices of the conservation of biological diversity abound among our people still. The levels of this conservation action are both at the individual and the collective. The underlying ethos in maintaining a sound bio–system in our region is one of responsibility and respect. Even the different ethnic identities of the people in Manipur were shown through the medium of Nature and the natural Way of life. The worship and reverence paid to inanimate objects along with living beings further extends this deep and conscious understanding of biological diversity at the community level.

Present trends and systems of education have led to a cultural myopia as they are now unable to see the consequences of their actions, the impact on their own children and grand–children. The human race can be said to be hurtling towards total colonization of Mother Earth, misguided by a notion of sole rights. Every thing around us, even the whole planet itself is being viewed today as a resource for humans and their enterprise. This anthropocentricism has led us further away from our own held and inner–most human values. Vast green covered zones and belts across the globe have turned into microscopic areas. Increasing advancing zones of desert and arid lands, flash floods, soil wash, soil erosion and siltation of our wetlands, cyclical droughts that have become almost perennial and growing areas of micro–climatic changes are the harvest of this mindless and wanton colonization of Mother Earth.

We, as humans and living partners among thousands of other living beings on this planet, have also begun to grow in numbers that are not only unsustainable but destructive to ourselves and our environment. We are definitely losing our way and our heritage. It is urgent that we turn once more towards our ancient and sophisticated indigenous sciences without losing a further second. We must now regain our lost knowledge and wisdom that have done us proud through the ages. For our lessons, we have to now walk to the lands and villages of the indigenous people who are being isolated and disaffirm, but who are the only true trustees and guardians of our biological diversity.

At the international level, clearly, great vigil needs to be kept for even these last vestiges of our biological diversity and our practices are in danger of being usurped through the use of powerful forces and mechanisms. Just as it is important to learn from our indigenous people, it is equally vital for us to be kept abreast of happenings and threats to our heritage. Meaningful participation and consultation of indigenous people in the measures to be developed and implementation of the Convention on Biological Diversity is our right which must be recognised and respected. This is indeed the very last chance for Mother Earth and all of us her children.

2

Preservation of Biotic Resources

Introduction

The gifts of Nature and human development cannot `be separated from each other. Thus we have to study not only the distribution of natural resources on the earth's surface but also how this distribution has been utilized by man for his material progress. Ecology tells us that the biosphere is unitary, that the natural life support systems of the species on earth are highly interlinked. The preservation of the delicately balanced ecological system is in the interest of the whole mankind. Here in this context we may recall the words of Radha Kamal Mukherjee, "the human group must be considered in a stable equilibrium not merely with reference to temperature, humidity, sunshine, altitude etc., but also to their indirect effects, the interwoven chain of biotic communities to which it is inextricably linked, the plants it cultivates, the animals to breed and even the insects which are indigeneous in the region."

Environmental scientists of the world are profoundly concerned over man's abuse of nature in the name of human progress. Under the influence of the stimuli provided by Nature, or as a result of the efforts man makes for satisfying his wants, the material progress is created. Land is cultivated, irrigation is provided, industries are developed, rivers are bridged, roads and rail networks are built, towns are grown and there is the whole paraphernalia of modern civilization which is born simply out of man's efforts for satisfying his want. He ruthlessly destroyed the natural resources of the world. Environmental scientists warn that if the present rate of air, soil and water pollution persisted the biological rhythm on the earth would be irreversely upset and man himself might face a long drawn-out process of extinction. The industrial and agricultural progress which bring a higher standard of living also have undesirable effects on the physical environment of man.

We live in an environment which is created by biological activities. The oxygen we breathe is produced by green plants by using carbon dioxide and so that the environment is a self-cleaning system. There are also systems in the soil which enable it to absorb human waste and convert it into humus which is then able to support the growth of plants which in turn feed people. It is also essential to be mentioned here that the process of photosynthesis forms the primary pathway by which carbon is withdrawn from carbon-dioxide "pool" and is used by plants to build carbohydrates and other organic compounds. These compounds transfer the carbon to herbivores which eat the plants. When herbivores are eaten by carnivores, the carbon moves further along the food-chain. Both plants and animals extract energy from these organic compounds by the complex biochemical process called cellular respiration. So there is a cycle of ecological process. Now-a-days in some of the advanced and developing countries, like U.S.A. and India the cycle is broken by dumping the human waste into water where it integrates with the water cycle, even after treatment. Moreover, it is not returned to the soil cycle. These circumstances are further aggravated by the application of modern inputs like chemical fertilizers and pesticide and insecticide etc. This breaks the ecological integrity of both the soil cycle and water cycle. This is an example of the way in which we are beginning to break down the self contained, self regulated properties of the environment.

Population Explosion and Natural Resources

It is generally believed that in future it will be impossible to satisfy the demands of the ever increasing population of the world from natural resources alone. The question arises whether their apprehensions are justified and if so, to what extent. To answer this question it is necessary to consider to estimate growth in the population and possibilities of providing food accommodation and waste as well as the possibilities of supplying adequate mineral and power resources for future demands. Many great thinkers and scientists of present century assert that in the near future the globe will be covered with cities from one pole to the other without agriculture and mankind will have to depend on synthetic food.

At the beginning of human history estimates of world

population is vague but estimation of the early 7000 B.C. was from 5 to 20 million. At the beginning of the Christian era, this figure had increased to 100 to 300 million. Fifteen hundred years later, in the time of Columbus, it was approximately 500 million.

In 1700, the figure had reached 700 million. It is estimated that by the year 2000 A.D. i.e. only 2 years from now onward the figure will be of 6424 million if the present growth rate continues.

As the population soars, productive land shrinks. Because the land area is fixed and inextensible while the population is increasing.

Biotic Resources

In the light of the above statement we may study the importance of the preservation of biotic resources. Biotic resources are often classified into two major groups, viz., Animal resources and Forest resources.

Animals and plants live in communities. Organism in communities may live together because of common tolerances for the non-living environment or they may be unified by the interaction between the organism themselves.

Animal resources and plant resources are closely related, alike in their fundamental characters of both classes or kingdoms. The actual boundaries between animals and plants are artificial and are based on ecosystem. The animal and the plant alike require food to repair waste, to build up new tissue and to provide material which, by chemical change, may liberate the energy which appears in the processes of life. Both animals and plants take their water and inorganic salts directly as such. The animal cell can absorb its carbohydrate and protiens food only in the form of complex organic substances : it is dependent, in fact, on the pre-existence of these organic substances, the products of living matter, and in this respect the animal is essentially a parasite on existing animal and plant life. The plant, on the other hand, if to be green, containing chlorophyll, is capable, in the presence of light, building up of both carabohydrate material and protein material from inorganic salts.

An important type of relationship within a community is in

food chains and food cycles. This cycle is the total system of energy exchange of a biotic community, usually including a complex of interlocking food chain. A particular link in a given type of food chain is often called an ecological niche. The ecological niche, on the other hand, is the term that includes not only the physical space occupied by an organism, but also its functional role in the community. The food supply of one animal is often the by-product of the activity of another. The relations between communities are reciprocal. Thus the relations between species in a community are never entirely one-way.

In ecological parlance, any given communities are the product of climatic complex. In climatic complex many kinds of organisms are always associated together and their relationships between the biological community and the inorganic environment are reciprocal. The type of forest depends on the type of soil, but the soil type is the product of the forest. The temperature, humidity and light conditions within the forest are the result of the vegetation, yet this forest climate in turn determines the type of community that inhabits the forest. Therefore, to maintain a balance eco-system is the utmost necessity of humankind.

A Geo-ecological Feature of Manipur

The Natural vegetation of the state bears evidence of the topographic and climatic changes that took place in the past. Studies of the administrative reports and accounts of the Pre-British and British periods reveal that the area was once an ideal home for many endemic species of flora and fauna. It is also noted that the area was once covered by a thick evergreen canopy of vegetation.

The natural vegetation of the state, as that of any mountainous region within the monsoon belt, varies with changes in the topographic feature, altitudes and soils. Thus, varieties of flora varying from the aquatic to those growing on high altitudes are found here. In the hills there are 5(five) major types of forests. This classification is sharply stratified by altitude. The forest types are (a) Wet Temperate Forest, (b) Pine-forest, (c) Wet sub-tropical Hill forest, (d) Semi Evergreen Forest and (e) Teak-gurjam Forest. And in the valley region there are 4 (four) vegetational zones viz., (a) Mixed open scrub and scattered tree, (b) Swamp-vegetation,

(c) Grasses and herbs and (d) Mixed reverine forest.

Hill Zone

(a) Wet Temperate Forest - These forests are localised at an elevation of 1,700 to 2,700 metres above sea level. The species are Quercus lemelloss, Q limesta; Michelia, Acer Mangnolia etc. and covers an area of about 1450 km^2.

(b) Pine forest - This type of forest is found in the north-eastern and southern parts of the state. It lies at an average altitude of 900 to 1800 meters and covers an area of about 2,440 km^2. Most of the species are of pinus Krasiya along with Quercus and Castanopsis.

(c) Wet Sub-Tropical Hill Forest - It stretches at an elevation of about 900 to 1800 metres a.m.s.I. covering an area of about 6590 km^2. Predominant species are Machilus, Schima, Wallichii, Quercus etc.

(d) Semi-Evergreen Forest - From the foot-hill zones upto 900 metres a.m.s.I. this type of forest is found. This type is gregariously grown in the western part of the state adjoining Cachar District of Assam. Artocarpus charplasha, Phoebe lanceolata, Toona calista, Gmeline, Adina, Schima, Wallichii, etc. are the common species of this region. This region covers an area of about 645 km^2 and have been subjected to heavy jhumming practices.

(e) Teak Gurjan Forest - This type of forest is commonly found along the border zone of Indo-Burma at the same altitudinal zone as the semi-evergreen forest is. It covers an area of about 510 km^2.

Valley Region

(1) Mixed opened scrub and scattered trees - This type of vegetation occupy the foot-hill and hillocks at the valley. It covers an area of about 1009 km^2. Some of important species are Calophyllum, Castanopsis, indica, phoebehensiana, Bombax malabarcicum, pine and Thea sinensis etc.

(2) Swamp Vegetation - The aquatic and swamp vegetation abound the lakes, swamps and marshes which covers an area of about 434 km^2. Water hyacinth, water lily and many species of

plants belonging to the family of Nymphaeaceae, Euryale and Nelumlum nucifira are found in this group. Certain species like Trapabirpinesa (Heikak), Cyperus ratundus (Kaothum), Segitaria sagittifolia (Kaokha), Zinania latifolia (Eshing Kambong) and under-ground rhizomes of lotus are used as food. However, water hyacinth has, of late, invaded almost all the lakes and swamps of the valley to the peril of indigenous aquatic flora and fauna. It is also interesting to note that some of the insectiverous plants like BLADER WART (Utriculsia) grow in the marshy area of the Loktak lake. Some of the swamps have now been reclaimed and encroached for cultivation and thus some of the species of aquatic plants are in the way of extinction.

(3) Grasses and Herbs - The areas around the lakes and marshes and the marginal plain bordering the foot-hills are covered with grasses and herbs. It covers an area of about 30 km^2. Reeds, tall grasses and various medicinal and edible herbs including Polygonum barbatum (Yellang), Plumbagozuilanica (Kengngoi), Centella asiatica (Peruk) etc. are some of the common species of this group of vegetable.

(4) Riverine forest of mixed type - Most of the river banks, flood-plains and the area around the villages support a type of mixed evergreen and deciduous vegetation containing Bambusa, Ficus bengalnes, Manggfera indica etc. It covers an area of about 27 km^2.

Fauna

Though the State enjoy a fair degree of uniform climatic condition, it has varying physiographic units clad with varieties of flora as mentioned earlier. Similarly the State supports a variety of faunal species. The Phyla of animals are the part and parcel of the forest complex and water bodies of the State. Thus, any form of destruction in the plant kingdom has a corresponding effect on the faunal kingdom.

The continuous increase of human population in the State, its activities in the form of deforestation of the land surface daming of the rivers, etc. have caused imbalance in the eco-system of the region. However, the available faunal species of the state are mentioned groupwise below.

Like any other part of the monsoon India, the common invertebrates found in th e State belong to the following classes :

(i) Phylum - Arthopoda which consists of common insects like common fly, mosquito, economic and uneconomic bee, butterfly, grasshopper, shrimps, prawn, mantis, etc. (ii) Phylum Annelida, including earthworm, leech, *milliped, centipede* etc. and (iii) Phylum - Mollusca includes various sub-species of snails, oyster, etc.

The southern portion of the Manipur valley is occupied by about 23 shallow lakes and swamps. Fish is abundant in these areas. Some of the rare varieties of fish of the valley are Dareng (Wallago attu), Porom (Channa stritus), Boggra (C. punctatus), Ngakra (Clarias batrachus), Ukabi (Anabas testudineus), Ngapemma (Colisa fasciatus), Ngatol (Labeo pungusia), Pengba (Otesbrama belangeri), Ngara (Tortor) and Ngakha (Puntius ticto) etc. Some of these species such as Ngara, Nganoi etc. are on the brink of extinction, because of growing ecological imbalance and introduction of carp fisheries.

Among the amphibians endemic ones of this region are those belonging to Hylidae and Pelebatidae (Frog & Toads) groups. Among the reptiles, Boa constrictus and large size Python are found in the swamps near the foot-hills. Some of the common poisonous snakes are also found in this region. Skink and other common lizards are also found in this area.

The avi-faunal species of this region are those common to other parts of the country. However, varied game birds that seasonally migrate nestle in the marshes and lakes of the valley. Among these species like gray leg goose, Burmese pea fowl, etc. are the main. With the depletion of the forests the large carnivores like leopard and tiger, once common in the state are on the verge of extinction. Other mammalian groups like moneky, procupine, mole, rat and mouse are also common in this area. One of the most rare species of mammals, brow antlered deer, locally known as SANGAI is found here as a localised species of the swamps of the valley.

Conclusion

Some specific measures have to be taken up in carrying out

State's aforestation programmes. State's forest department is progressing in the aforestation programme. Mostly exotic varieties are introduced in this programme without observing enough quarantine period on these varieties. Such plantation programme being carried out, shall react to the local climatic condition of the region and shall also change the forest complex of the State.

References

1. Forest Survey of India Vegetation Map of Manipur based on Landsat Imagery, 1980.

2. Land-Use Pattern in the Manipur Valley : A Doctoral Thesis of Gauhati University, 1981.

3

People, Space and Economy of the Manipur Valley

A study on FUTUROLOGY of this part of the country has to be partly theoretical and partly empirical with greater emphasis on the latter. There are several theories regarding the planning and forecasting for future based on the present existing indices. This is a geographical study in the sense that the problems are viewed from the spatial angle and from the point of land use and linkage of the growth of population in the area. The gradual expansion or defusion of urban phenomenon to the arial extent of the arable or culturable land and the spatial distribution of population within the Manipur Valley are the main theme of this study. The study also focuses attention on the physiographic factors which influence the internal structure and external projections of the land-use pattern of the valley.

This study is planning oriented and gives guidelines on the planning, designed to meet the demands of the ever-increasing population at present as well as in the Twenty-first century. It also provides necessary guidelines for planners to do the comprehensive physical and social planning of the Manipur Valley.

Empirical analyses of this type attempted have been made in the past in India by several geographers, economists and sociologists, focusing attention on different aspects of planning. Several geographers have, however, tried to integrate the concepts of development in different disciplines and to suggest a theoretical framework which would provide solutions to the land-use problems both physical and social.

Planning for the region (Manipur Valley) has to be both clinical and preventive. As a clinical endeavour, it is designed to solve the existing problems. It would look at the valley area from various angles and suggest, after proper and adequate diagnosis, the steps to be taken to put the system in order.

To an amateur planner the two aspects of planning may appear to involve separate policies and hence needing different approaches. This however, is not true. Both for clinical and preventive purposes the planner has to examine the problems that the valley faces and has to understand the forces-geographic, economic, social and others which have brought the valley area to the level that it is against the development of the Manipur Valley in the 21st century. He has to understand the institutions and the limitations within which he himself has to function and also know the nature of services the community needs today and will need in future.

Most of the developmental planning organisations deal only with physical aspects of planning. This is especially so in India. But we know that physical planning can succeed only if it is based on realistic assessment of the socio-economic problems and issues involved in the land-use planning of the valley. A planner must get sufficient data on the present and the future needs of the valley before he can prepare a good physical plan. He must know the present and future problems associated with housing, work, movement, recreation and education, before he can even think of a physical plan. Above all he must know the socio-economic needs of the valley before he can think of an effective and meaningful use of land.

To get the data needed for the plans, all planning endeavours have to be backed by realistic diagnostic surveys by specialists in various fields. Absence of such studies makes a physical plan nothing more than an attempt to solve the problems before knowing what the problems etc are. If planning efforts have so far failed to raise the economy of the Manipur Valley, it may be because the efforts were not based on sound diagnosis of the problems. This study is designed to fill this gap.

People

The total population of the Valley according to the census of 1991 is 14,77,760. This figure was reached through a rapid growth of population during the recent decades.

Consequent upon continued Burmese invasion, from 1755

A.D., a large number of inhabitants left the valley and it was estimated that there were only 2,000 persons in it in 1824 A.D., a miserable remnant of a thriving population that existed before the Burmese invasion in the later half of the 18th and the beginning of the 19th centuries (Bhattacharyya, M : 1963). The population of the State increased rapidly thereafter and according to Brown, its total population in 1873 was 1,39,000 out of which 65,000 was on the Valley (Brown, R., 1873). The first regular census of the State was carried out in 1891 which put the total population of the State at 2,21,070 (Allen, B.C., 1905). Out of that the valley had 1,35,288 persons and the density of population was 52 persons/km^2.

The growth of population during the period 1901-1971 in the valley was spectacular (table 1:1). The decadal percentage increase was between 11.64 to 19.9 during the first half of the century but it shot up to 29.51 per cent in the decade 1951-61 and 39.92 per cent in the decade 1961-71. The sudden increase in the rate of growth after 1950 was mainly *due to removal of State's ban on immigration and decline of the death rate.* The 1951 census recorded 2,765 immigrants of whom 1,687 were from Nepal, 608 from the then East Pakistan (now Bangladesh); 304 from elsewhere outside India and 166 persons from other states of India. By 1971 census the number of immigrants rose to 35,124 of which 11,580 persons were foreigners, mainly Nepalis and Bangladeshis and 19,544 persons were from the States of India outside Manipur. Subsequently, in 1976 a "working paper on the population Growth of the next Decade", of the Secretariat Planning, Government of Manipur (1976) recorded 95,674 immigrants in the Manipur Valley and 1,42,033 immigrants in the State as a whole.

Although the total geographical area of the valley is only 2,639.8 km^2 or 12 per cent of the State's total area (22,347 km^2) it contains 69 per cent of the State's population (1971). The population of the valley consists of 80 per cent of Meiteis or Manipuri Hindus, 10 per cent of Manipuri Muslims, 7 per cent of Scheduled tribes and 3 per cent of Scheduled Castes and the immigrants of all castes and creeds. The very low density (17 persons/km^2) of population in the hills of Manipur is mainly due to rugged topography, made up of steep slopes, forests and consequent lack of agricultural opportunities, mineral and other resources having remained unexplored so far. The lack of suitable transport system has added to the adverse economic condition of the hills.

According to 1981 census, the arithmetic density of the valley is 300 persons/km^2, which is much higher than the all India average of 167. However, this figure does not give the exact picture of human density in the valley. Because, out of the total valley area about 546 km^2 is occupied by lakes, swamps, marshes, barren uplands and hillocks. Eliminating the negative area covered by these features, the density of the valley is calculated at 400 persons/km^2. The available arable land of the valley is only 961 km^2 or 36.4 per cent of the valley as a whole with an agricultural density of about 145 persons/km^2 (1981).

The arable area of the valley is only 96,046.153 hectare which accounts for only 0.13 hectare per head. Hence, eight persons have to depend on a hectare. The present problem is further aggravated by the high growth rate (3.4 per cent per annum) of the population of the valley. It is estimated that the projected* population of the valley in 2001 A.D. will be approximately 2.08 millions of people (table 1:2). Thus it is expected that with the population raising at the present rate, the pressure on land will be tremendous.

On the total population of the valley, 62.4 per cent are settled in the rural areas and 60.3 per cent of the working force are directly engaged in agriculture. The rural density rose to 233 persons/km^2 in the year 1971 with a growth rate of 39.9 per cent. Between 1951 and 1971, urban population also grew rapidly and considerable rural areas were included in the fold of the towns (table 1:3). Hence, by the process of urbanisation diminishes about 0.15 per cent of the rural areas annually.

The agricultural density of the valley also rose from 132

---

* Population Projection Formula

$PX = Po\left(1+\frac{r}{100}\right)^n$, where PX = population of the area under projection, r = average rate of growth, n = number of years for which the population is to be projected, Po = Population of the base period.

person/km^2 in 1961 to 141 persons/km^2 in 1971. The economic division of the population of the valley reveals that while working population constitutes 29.9 per cent, the non-working population accounts for more than two-thirds putting an immensely heavy burden on the former. This section includes the population belonging to the age groups of 0 to 14 years and above. Moreover, about 80 per cent of the State's total student population are concentrated in the valley itself. Above all more than 99,671 persons under the age group of 15-24 years have been enumerated as dependent. They include educated unemployed and those undergoing higher studies.

Working Population

Out of the total working population of the valley about 60.34 per cent is directly related to agriculture. It represents 18 per cent of the total population of the valley which composed of 89.6 per cent agricultural labourers apart from 1.6 per cent engaged in livestock, forestry, fishing, plantation, orchards, etc. The agricultural working force is obviously dominated by male while in the household industries the female participation is spectacular. It is, however, important to note that the proportion of population engaged in the household industries decreased by 50 per cent between 1961, 1971 and 1981. This is because less profit accrues in the household (handloom) industries. Abandonment of household industries means an increasing pressure in the agricultural sector as there is little scope in other occupations. There has been a spectacular increase in the group, agricultural labourers, during the mentioned period (from 1,137 in 1961 to 11,808 persons in 1971). This increase has an impact on the land-use of the valley.

Space Problem

Various limitations and impediments in the utilisation of the available land resources of the valley are inherent in its physiography, soil conditions, climatic characteristics and socio-cultural milieu. Besides, the increasing pressure of population on the agricultural land, excessive subdivision and fragmentation of the holdings have reduced the per capita available land for agriculture in the valley to very small parcels. This situation has reflected itself in very low intensity in respect of agricultural land-use and cropping, low production and insignificant land-use development. On the other hand, population growth in the valley is creating a serious problem of unemployment among the peasants.

The increasing population needing more and more land for both agricultural and non-agricultural uses has posed a serious problem in the land-use planning of the valley. There are two ways of which agricultural output can be raised, firstly, by extending the area under cultivation and secondly by raising the output per hectare.

The increase and absorption of cultivated area of the valley after 1951 can be observed into two phases, namely pre-1970 and post-1970 period.

Most of the additional arable land reclaimed between 1951 and 1971 were in Serou, Saiton, Kwakta, Moirangkhunou, Keirenphabi, Komla Khong, Mayang Imphal, Kanglatombi, etc. where the new settlements of the Meitei, Manipuri Muslims and immigrants from Bangladesh and Nepal grew up. There was again a substantial addition to the cultivated hectarage after 1971 as a result of extension of the irrigational facilities. Thus, during 25 years from 1952 to 1977, there was an increase of 18.7 per cent of the net area sown. Correspondingly, the increase in the output of paddy in the valley was from 1,45,138 quintals to 2,12,252 quintals. However, productivity analysis shows that the average yield per hectare in the valley decreased from 2,026 kg/hectare in 1952-53 to 1,625 kg/hectare in 1977-78. Thus, it is clear that the increase in output came from expansion of the cultivated area rather than from any increase in the yield. The prospect of further increasing the net sown area is extremely limited, for almost all the arable land of the valley has already been brought under the plough. The areas under both forest and permanent pasture are decreasing fast. It has been mentioned that the per capita cultivated land decreased from 0.325 hectare in 1951 to 0.125 hectare in 1971. It has further decreased to 0.08 hectare in the year 1976.

It was also mentioned that a higher increase of population (39.92) is recorded during the decades 1961-71 and 1971-81, while it was only 19.9 per cent during 1941-51. Based on the FAO's recommendation of 0.5 hectares of cultivable land per person, the present population of the valley required extra arable land over and above the existing arable area of 72,306 hectares (1976). If it is calculated for the projected population of 21st century the required extra arable land will be 9,41,998 hectares. It is obvious that no expansion of arable land is feasible. In fact, the scarcity of land has been further aggravated by the sprawling of towns and associated infra-structural developments which have been eating up 0.15 per cent of the arable land annually. However, the only possible way of increasing the area under plough to some extent, is to reclaim the culturable waste, utilize the area left fallow and practice double and multiple cropping. The cost of reclaiming new land is hardly justifiable socio-environmentally, however economically desirable may it be. Nor does the expansion of the cultivated hectarage, as such, contributes to agricultural productivity. Thus, new plans for agricultural development should concentrate on intensifying the output within the existing arable area of the valley.

It is found that more than 98.5 per cent of the total cultivated area of the valley depends on rain for water supply. Most of the rain water during the monsoon goes away as run-off and is not preserved. Only a little portion of the rain water could be used by the peasants in the main paddy season. During the drought as also in the dry seasons, most of the area under plough is seriously affected, hindering double cropping. In fact, only 2.3 per cent of the total cultivated area of the valley was under more than one crop in the year 1976-77. It is, thus, obvious that lack of assured water supply pose an acute problem on the agricultural land-use.

Nearly 99 per cent of the peasants of the valley work on their land with a pair of bullocks and a wooden plough. Thus agriculture in the valley is carried out at a level even below the bed rock the present day mechanization of agriculture. Traditional implements and animal power are still widely used in the valley. Poverty and ignorance, coupled with poor extension services have served to keep modern agricultural implements out of the farmer reach. The peasant has often to borrow or hire even the implements which he knows how to use. Oxen and buffaloes have still remain as the main means of plough. Thus it is essential to presrve adequate grazing land. But acute population pressure accounts the low hectarage devoted to fodder crops. Nor do the livestock have any satisfactory alternative source for proper feeding. Thus so long as animal remain the source of plough power, the progress of double cropping in the valley will remain insignficant.

Conclusion

With an enormous pressure of population on the agricultural resources as expressed through an alarmingly high cultivated area ratio and with the farthest limits reached by the extent of cultivation, further expansion of area under plough appears limited. It would seem imperative that the cultivators of the valley should raise more than the existing quantity of the harvest from the same field to meet the need of the increasing population. Therefore, serious efforts should be made for intensive use of whatever land is available in the valley. The arable land in the valley has been increasingly used for growing foodgrains, mainly, paddy in response to the increasing food requirements indicating little commercialization of crops. Unless the yield per-unit area of paddy is increased, its hectarage cannot be reduced to make room for growing other cash-crops which would derive higher income for cultivators. Measures suggested to meet the exigency are transformation of the lingering traditional farm technology, supported by the essential infrastructure.

Intensification of land-use particularly that in agriculture, should be achieved not due to population pressure, but by a choice of scientific farming.

Industrializaion and development of tertiary sectors are so insignificant in the state that they have not been able to divert the surplus man-power from agriculture to other occupation. Therefore, urgent steps should be taken to develop the secondary and tertiary sectors of the industry in order to make the agricultural land-use of the valley economically more efficient by withdrawing the excess man-power from it.

Agriculture is the mainstay of the economy of the valley. The malady-ridden traditional socio-economic structure of the peasants must be rationalized and reorganised into a new pattern to fit into the productivity oriented economy of the modern world. Earnest efforts should be made for providing modern agricultural infra-structure, which includes among others, rural, electrification, agricultural credit, transport and communication etc., for optimum land-use.

The rate of literacy among the agricultural population of the valley is low. It is one of the most important factors of agricultural efficiencies, the latter should be ammended to prevent further decrease in farm size. Besides, new laws should be framed and introduced to debar the noncultivating adult members of a family employed in other occupation from inheriting and purchasing agricultural land.

Little attention to soil conservation has resulted in a great misuse and abuse of land in some parts of the valley. The land-use pattern needs re-orientation with a feasible and dynamic plan for the optimum use of every hectare of land. The paucity of land under forest is now realised. It is, therefore, suggested that the area under forest be increased at least to 30 per cent of the total area of the valley, particularly in the foothill and catchment regions and on the isolated hillocks, by afforestation.

Table 1.1

Population and its growths in the Manipur Valley (1901-1991)

Year	Population	Growth
1901	.1,89,744	—
1911	2,21,431	16.70
1921	2,59,614	17.24
1931	2,89,843	16.64
1941	3,43,694	18.58
1951	4,12,115	19.91
1961	5,33,732	29.51
1971	7,46,776	39.92
1981	10,50,460	40.66
1991	14,77,760	40.67

Source : Census of India, 1901-1991

Table 1.2

Projected Population of the Valley upto 2001

Year	Projected Population Figure
1991	14,77,760
2001	20,78,890

Table 1.3

Growth of Rural and Urban Population (1901-1991)

	RURAL		URBAN	
Year	Population	Growth Rate	Population	Growth Rate
1901	1,17,510	–	72,234	–
1911	1,46,781	24.90	74,650	3.34
1921	1,79,611	22.36	80,003	7.17
1931	2,04,039	13.60	85,804	7.25
1941	2,43,978	19.57	99,716	16.21
1951*	4,09,253	67.74	2,862	97.13
1961**	4,66,015	13.86	67,717	2,226.07
1971	6,05,284	29.88	1,41,492	108.95
1981			3,75,460	165.36
1991	1,14,183		4,98,311	34.67

Sources : Census of India, 1901-1991.

* Urban area was limited to town found area of 5.17 km^2.

** Urban area again extended to 17.48 km^2 and inclusion of new small towns area.

4

Land Use, Social Constraints and Development

After the First World War, an extensive reform in the agrarian structures of Central and South-Eastern countries of Europe has been made. As a result, feudalism was liquidated and peasant proprietorship was introduced. Even conservative England could not stop the emerging trend and had to enact legislation to split up big estates and create small holdings by passing Land Settlement Act, 1919 and Land Utilization Act, 1933. However, at the same time, collectivisation in USSR, on the other hand, has stood the test of time to a fair extent and has spread to other lands. But after the Second World War, when democracy and political freedom were achieved by Asian countries, the issues became still more confusing. A via media is being tried and co-operative farming has taken the field, as an agent of reconciliation between the two extremes of views. In order to have a dynamic land system, we need at least two fundaments i.e. (i) to produce efficiently and (ii) to be in tune with the sociological values of the country or the State. Sometimes one of these has to be sacrificed and more importance attached to the other. The relative efficiency of various systems of farming viz., collective farming, land-lordism or peasant proprietorships, can be ascertained with comparative case studies with a little practical research. But the problem becomes more complicated when we remember that agricultural land use not only provides food and raw materials but also is a means for social organization and is one of the most potent contents of any civilization. Land tenure, which is the common denominator of agricultural organization, everywhere, must fit into the prevalent pattern of culture in the country, whether it is based on capitalistic motives of individual effort or is directed towards realisation of joint effort in a mass organised society.

Our understanding of the complex reality depends on the questions we ask; these in turn depend upon the culture into which we have been socialized and the viewpoint or discipline in which we have been trained. A fact is only a fact in terms of a theory and problems are defined as such within an implicit or explicit theoretical framework. That such conceptual schemes are more often only tacitly accepted in empirical investigation should be a matter of deep concern to social scientists. It seems clear that any attempt to explain the ordering and pattern of human activity is constricted, if not forced, into a monocausal framework.

Geographers in particular, who may attempt to explain medieval field-systems or economic growth in modernising countries by the facts, should see the need for a more rigorous theoretical framework. Seemingly every generation is obliged to rewrite the past in terms of its own values and ideologies. Yet, however strongly we may react against a naive historicism, it cannot be denied that an understanding of the past is internalized in the individual-particularly in certain charismatic leaders - and may be partly responsible for spatial and temporal diversity of organisation of human activity.

Based on the point of view reflected above, the present paper is trying to focus on all the major social constraints exerted upon the landuse pattern of both the valley and the hilly terrains of Manipur. Here, in this context, the author is examining the social constraints mainly from the angle of the land holding and its ownership and tenure systems in the State in relation to its optimal utilisation. Because, it is believed that the size of holdings, ownerships and tenant systems have a more direct relationship with the economic prosperity of the available land lies in the size of the holdings and not in land tenure. There is no doubt that the size of holdings has an intimate and direct bearing on the efficiency of land-use; but one should understand the fact that land tenures and rights over land are no less important.

Although many land laws like Manipur Land Reforms and Land Revenue Act, 1960 and other Ceiling Acts have been introduced in the State, there are still various social customary laws in vogue. Tenants with heritable rights in the valley and possession right in the Hills may be mentioned in this context. These customary and social practices are, however, found to be worse in the valley arable land. Simultaneously, we find extreme shortening of Jhum cycles in the Hills. This has been further aggravated by the Laws of Inheritance which provide equal share to all the male descendants. As a result the number of holdings and fragmentation of holdings in both directions has significantly increased. Thus, Manipur valley has become a land, small and uneconomic in size i.e. below 1.5 hectares. This holding size is far below the all India average of 2.3 hectares. The agricultural efficiency under such conditions becomes less. The technology by which these small plots are cultivated is primitive and therefore the yield is very low. Small farmers grow low price crops mainly owing to unavailability of inputs or capacity to invest.

The poor returns that he gets can hardly enthuse him to avoid negligence to the land. Thus the small plots become neglected and economically unviable.

In a State like Manipur where the peasant-holding system has been prevalent for hundreds of years, the available productive land per capita is bound to decrease with the increase in population. As there is little scope for livelihood in other sectors of economy, each generation enters into patrimony and the extent of land which goes to the share of each of them gets diminished progressively. As a result, the operational holdings of each nuclear family is of a size below one hectare. Almost 90% of the total holdings in the valley constitute small and marginal farming unit holdings. It has also been alarmingly reported that most of Jhuming areas of the hills have become almost barren and economically unproductive. This is further aggravated by the extreme shortening of Jhum cycles (2-3 years' cycle).

On the whole, the problem of population growth coupled with the socio-cultural phenomenon have re-acted seriously on the utilization of land resources. Development in time and development in space are constantly linked together. Therefore, it is high time to diagnose the problems for prescribing a new economic and cultural base and removing the constraints.

Let us take a concrete example. The diagram below represents the traditional land use system among the Tangkhuls in Manipur.

1. Settlement area - (Homestead land)
2. Individual/household plot (tree & vegetable cultivation)

3. Clanial land (preserving good varieties of wood and trees)

4. Village/community/public land (Jhuming).

Socio-culturally, Tangkhul land has been classified as mentioned in the diagram above. The first and core one is used only for construction of dwelling houses and domestic cattle shed. The second circle is distributed to each and every individual household for any kind of activities which is related to his livelihood. The third one belongs to the head of the clan. However, individual member of the clan may utilise the plot with due permission from the head of the clan. The outer most circle (4th) belongs to the village where individual households may practice Jhum and other economic activities etc. Thus, there are definite Tangkhul sociological factors controlling the use of available land. However, these customary right have been rapidly vanishing from the society and are being more and more dominated by the outer most circle. Therefore, while analysing the problems with respect to the prevailing environmental factors, one has to choose a system so that the tapping of land resources in the area may be more fruitful. Considering the sociological aspect of the Jhuming cultivation and also the present trend of environmental degradation, we find serious constraints for maximal land resource utilization. We may think of transforming the Jhum land into the modern collective farming. Simultaneously we may think of land consolidation programmes in the valley. It is also essential to conduct a systematic land use and Land potential survey in the state.

It may have been realized from the above discussion that the problems are of a serious and urgent nature. They require the immediate attention of all who feel concerned about developments in Manipur. It might also have become clear that the problems are complex and therefore no simple solution can be suggested. I, therefore, venture to propose a model which may be worked upon to prevent this economically backward state from further deprivation.

We have pointed out earlier that the old system of land use was suitable for olden times but has now disintegrated due to population growth and other factors. Unfortunately, no new system has been developed to take care of optimal land use as well as ecological necessities. In the context of this vacuum, we may try to develop for each specific area an economically viable and socio-culturally acceptable system on the basis of the multi-dimensional model presented above.

Socio-Environmental Model

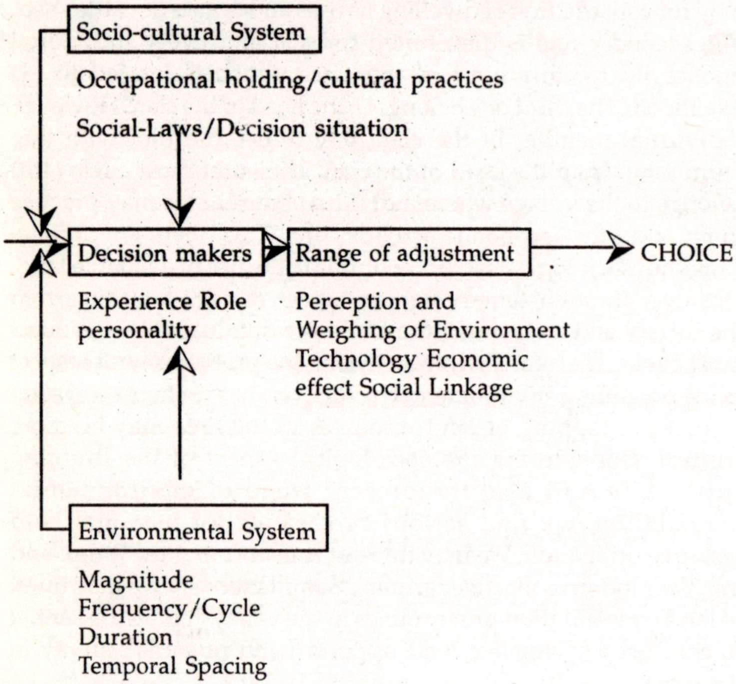

Existing Problems of Land-use

Physiographically, Manipur is a land of hill ranges and intermontane valleys. More than 80% of the State's total area is under the regime of terrain topography which is presumed to be the area of forest and forest dwellers. The remaining, less than 12% of the State's total area, is treated as valley with tremendous population pressure. The continuous growth of population and manifold expansion of economic activities along with rapid urbanisation throughout the hills and plains of Manipur, has affected the environmental conditions alarmingly. It is also a fact that whenever the growth and density of population increases, the productive land resources, including forested area, also shrink. Consequently, even the required minimum area to be kept under the coverage of green vegetation or forested area has been threatened at a faster rate. Ultimately, these processes lead towards the danger zone of Environmental degradation.

Today, we are facing the cyclic problems of flood, drought and siltation in the State simultaneously in the same year. Evidently, this is a direct resultant phenomenon of the degradation of our natural environment. These are the common problems caused by the degradation which is the action and reaction of all the activities of the people who settled in this State. In this context, our life styles, consumption patterns and reckless consumerism in the State is one of the root causes for the degradation of its environment. The present day life style is non-sustainable. We are consuming faster than the earth can replenish. We are consuming faster than the earth can assimilate. The progressive growth of technology and the resultant consumerism and waste have encroached deep into our societies. We do not seem to have realised the cost of destroying our environment. Do we realise the cost of a tree which cleans air, regulates water-flow, produces bio-mass & provides habitat to other constituents of flora & fauna? Can we develop forests which have taken thousands of years to build their eco-system? Can we stop the disturbance to hydrology of an area which regulates climatic conditions ? These are the most fundamental environmental concerns for all of us today.

The problems of Environmental degradation is further aggravated by the continuous growth of population on a severely limited land area and natural resources. The state has 22,327 sq.km. only as its geographical area whereas its population has increased as follows :-

Decadial Census	Total population	Growth rate	Geographical area
1901	2,84,465	–	23,327 sq.km.
1911	3,46,222	+ 21.71	-do-
1921	3,84,016	+ 10.92	-do-
1931	4,45,606	+ 16.04	-do-
1941	5,12,069	+ 14.92	-do-
1951	5,77,635	+ 12.80	-do-
1961	7,80,037	+ 35.04	-do-
1971	10,72,753	+ 37.52	-do-
1981	14,11,375	+ 32.46	-do-
1991	18,26,714	+ 28.56	-do-

Whenever there is a growth of population,. the demand for land for construction of houses, consumption of food, forest products and demand for other amenities also increase. To meet these demands one has to depend on the available resources without further thought for conservation and protection. Hence to multiply the production from the limited arable land the cultivator has to apply improved technology by using HYV crops with chemical fertilizers and chemical pesticides. Thus, ultimately polluting our own water systems and soils to such an extent that certain constitutents of nature and its resources are destroyed permanently and they cannot be rejuvinated to their original condition. These changing patterns of land use have already been observed in the valley of Manipur. The changing land use pattern in the valley has been further influenced by the trend of urbanisation. Expansion of urbanisation encroaches upon arable land at the rate of 3% annually. Thus, urbanisation is also at the cost of the arable land. On the other hand, urbanisation means increasing number of C.I. Sheet and R.C.C. roofing houses or areas and also metalling the road. As a result of this a huge accumulation of water during the monsoon are permitted to free flow as a surface run-off which sometimes creates flash floods in the city areas.

Again in the hilly region more than 900 sq.km. of forested area has been transformed into jhum field annually. Further by the increase of population growth and expansion of infrastructural facilities of urbanisation the jhum-cycle has been shortened from 20-30 years to 2-3 years. Over and above the trend of illegal logging business has also been increased tremendously. With all these factors clubbed together, today the Sate has only 3.22% of its total area under dense forest, while the rest is under degraded forest which is commonly termed as abandoned Jhum field and wasteland. Thus the area of wasteland reached upto 4,90,527 hectares in the year 1990-91. Depending on these factors if some one visible all the degraded barren land, recklessly exposed to the agents of the weathering waiting for denuding the slopes of the hills. These processes accelerate surface run-off, soilwash/ erosion, siltation in river bed and lakes in the valley causing flash flood and droughts alternatively.

On the other hand, Manipur is geographically located in a zone where the world's heaviest rainfall is recorded. At present, the State receives 142 cm rainfall annually. It receives rains during the summer monsoon from the middle of June through to September. During these four months, it receives 60% of the total annual rainfall. However, because of various environmental problems, particularly in the major catchment areas of the two major drainage systems viz., the Barak and the Manipur rivers are exclusively rainfed, with torrential headstreams. Most of the tributaries of both the drainage systems were perennial but now all of them are in lingering conditions. Most of the tributaries and the numerous fountains and springs have become seasonal. This occurrence of seasonality in the drainage system and siltation in river beds and lakes in the valley of the State has a significant correlationship with the increasing trend of deforestation in the catchment area. In fact, forests of the catchment area are the watershed areas where store the monsoon rain to feed the rivers all the year round. Presuming that we are all aware of these problems at different levels what are we doing to save the situation. Unfortunately, there are no conscientious efforts made by the academicians and bureaucratic officers to save the situation progressively. There may be some individuals who are trying their best; but such contributions are only a few drops in the ocean. What we need is large scale and concerted efforts to save mother nature to save ourselves. In order to maintain a better eco-system and better environment in a State like Manipur, it is necessary to maintain at least 33% of the State's total area with forest coverage and green vegetation (as per New forest policy of India). For this, we should keenly observe the ground realities, not what official records say. The crucial question is how to ensure the maintenance of this forest area ? Some scholars argue that it can be done if jhum cultivation is totally stopped. But then, how to stop jhumming? Jhumming is the culture of the tribal highlanders. To change the age-old practice without proper and adequate protection and without environment and occupational rehabilitation would be a suicidal proposition.

However, we also know that massive aforestation programmes have been taken up under social forestry programme through community level as well as non-governmental organisations. Sometimes we also observe that under large scale social forestry scheme to plant a tree and to protect a tree sapling 4 to 6 growing trees were cut for fencing. Thus it may be calculated that in Manipur the ratio of afforestation to deforestation is 1:18:1.

Secondly, all the social forestry programmes were implemented without proper land use planning. There is no selection of tree species. Just on a normal routine way NGOs and concerned sections have planted trees without seeing the natural suitability and without calculating future income generation. Here in Manipur, easy and quick growing trees like eucalyptus, acasia etc. are planted without understanding the repercussions. Why are we not planting species like teak, Sal and other high class timber quality trees in a long term perspective.

It would be better to plan the whole of this State social forestry programme after regionalising the zone with proper land use planning. In this context a region may be formed all along the catchment or watershed area as a region or micro-region for the development and growth of forest under social forestry programme. This will give better results in the long run. Lastly, the real solution lies in the actions of individuals, groups and communities all over the State. Corrective actions at village levels can only arrest the problems of environmental degradation and destruction. Only local, meaningful actions, not hollow rhetoric, can face the threat of environmental disaster.

A New Technique of Land-use in the Hill District of Manipur

The area of Manipur is 22,327 sq. kms. of which 2238 sq. kms. form the Manipur Valley and the remaining 90 percent of the total land area is hilly terrain. According to 1981 census, 4.9 lakhs of population are highlanders or so to say tribals inhabiting in the hilly region. The density of population in the hills is only 24 persons per sq. km. and that of the valley is 415 persons per sq. km. From this point of view, it is felt that the hilly regions of the State offer ample scope for more population, if only the available land area is scientifically utilized. 90 percent of the geographical area are inhabited by hill people substantially depending upon the remaining 10 percent of the State's area i.e. the Valley of Manipur for food production. While the fertile levelled ground of the valley provides ample opportunity for agricultural pursuits, the steep hill slopes with narrow and deep gorges, and sharp ridges of the hill districts leave little scope for profitable sedentary cultivation.

Therefore, the entire hilly regions of the State is not fit for the present system of agricultural land - use practices. The present system of land - use is totally based on shifting cultivation. It is also popularly known as JHUMMING and is considered as the first step in the transition from food - gathering or hunting culture to food production. This system of land - use is believed to have been originated in the Neolithic period around 7,000 B.C.[1]. This type of land - use is known as PAMLOU in Manipur. It is a cyclic cultivation in tropical and sub - tropical regions. The most primitive practise is still being in vogue through out the hilly tracts of Manipur. This system of land - use supports approximately 4.9 lakh Jhumias, which represent only 34.61 percent of the State's total population (1981). Continuance of this primitive practise have an adverse effect on the balance of the environmental eco - system of the region. Of course, such kind of land - use has a close link with socio - economic and cultural factors including lack of communication leading to physiographical remoteness and isolation. Therefore, it is also essential to study the Jhummimg system of the region.

The Present System of Jhumming :

The essential features of Jhumming as practised in the State are :

(i) Selection of site on hill slopes usually before December,

(ii) Clearing of site by cutting forests in December and January,

(iii) Burning of forests around mid - February to mid - March,

(iv) Planting of various crops in an intimate mixture by dibbling,

(v) Abandoning the land after cultivation for a period of two years usually and shifting to another site,

(vi) Return to same site after 3-6 years to repeat the same process.

No Drought power or improved equipment is deployed by jhummias for preparation of the land. The only implements used in farming are the Chopping knife, dibbling stick and a small hoe for weeding. Similarly no scientific technology is employed. The only inputs use are seeds and human labour. Except cutting of forests and watching to protect the crops, the other labour operations including weeding are usually carried out by the women. All essential crops, such as rice, maize, tapioca, colocasia, sesame, cururbits, beans, yams and banana, etc. are planted in an intimate mixture, although a single crop of rice is sometimes grown in the second year of jhumming. The land is left fallow for regeneration after 2 - 3 years of jhumming to be used again after some year. The jhumming cycle in the same plot of land which extended to 20-30 years in olden days has now been shortened to 3 - 6 years because of pressure on land due to increase in population and decrease in productivity leading to extension of more areas under jhumming. The area of fresh clearance of forest every year in Manipur is estimated to be 600 sq. kms. Jhumming accelerates surface run off, erosion, silting of dams and reservoirs and siltation in the river beds, causing flash flood in the rainy season and drought in other seasons[2]. Therefore, there are certain adverse effects of jhumming in the present eco - system of the region.

Although jhumming practice was good for olden days, yet it is considered, at present absolutely wasteful. The series of adverse effects caused by this system of land - use are summarized below:

There has been extensive deforestation causing large scale denudation of hill tops and slopes, thus leading to undesirable ecological changes and elimination of the watershed area of the region.

Extensive deforestation causes a large - scale erosion of soil and consequently producing adverse effects ; (a) silting of streams, river beds and marshy areas causing flood in plain areas and (b) loss of fertility which is not easily built up causing low productivity and subsequently pressure on land.

The continuance of primitive form of land - use offers very little scope for introduction of modern technology. This alongwith the loss of fertility has led to such low productivity that the jhummias live in near famine conditions.[3] For a particular period during January to July, they have to live on root crops, jack fruits, etc.

This system of land - use is labour - intensive with low technology, which results in low productivity. Further, there is no scope for development of sources of subsidiary income for jhummias. Besides these which are directly related to agriculture, there are other aspects which adversely affects social welfare ; the public health, education, communication and such other basic facilities are difficult to develop when there is no permanent settlement.

Jhumming Control and Inovation of New Techniques :

Each State or Union Territory of the NE Region has some programme of jhum control as special programme or as a part of the departmental programme of soil conservation and afforestation. Even then there are primary causes for the failure to attract the jhummias to permanent settlement as given below.

Introduction of new system of technique of land - use cut into their socio - cultural phenomenon abruptly. They are not used to cultivation in terrace using bullock/implements. At the initial stage of terracing, production is low as the top soil are removed during development. Water conservation practice etc. are also not properly developed for the region.

With all these factors *Schhippe* observed in his studies on shifting cultivation in Africa, "The modern civilization has failed to improve African agriculture because it has proved so far incompatible with the environment of the wet tropic ..... Agriculture is that sector of human activity in which there is greater inter - action between environment and human culture which has grown in and from it.[4]" He has, therefore, advocated improvement without doing violence to the limited frame work of tradition and environment. Majumdar also feels that shifting cultivation in the region cannot be avoided for a long time and hence conscientious efforts are needed to improve productivity of jhum cultivation and minimise soil erosion. While doing so he advocates. " An important aspect duly recognised as a crucial strategy for development is the need to tailor the programmes to the local conditions and values.[5]"

Unless the attitude of the jhummias is changed and they are convinced of the need for settled agriculture and terracing, the present governmental schemes will be hard put to the task. But, it would not be desirable to allow the soil and fertility erosion to continue for long years. Yet, there is no proven technology/ message with the development department to bring about the awareness amongst the jhummias, because there has not been systematic study of the problems faced by the jhummias, particularly with regard to improvement of agriculture in the environment in which they live. It may be of interest to note that there are slopes of upto 100 per cent or even more. In fact, in Ukhrul District, hills are so steep that areas with less than 50 per cent slope are very limited.

Soil loss, runoff, productivity of crops, etc. are compared in four systems of farming, namely jhumming (WI) partial (1/3rd at base), terracing (w2), complete bench terracing (W3) and puertorican terracing (W4). The plots are developed in the water - shed areas of Thoubal and Iril river of the State. In the partially terraced plot, field crops are grown only on the 1/3rd terraced plot. Horticultural crops are grown with crescent terraces on the upper 2/3 slope. Leguminous fodder namely cowpea is grown between the horticultural crops and the lower terraces to provide fodder, as well as to cover the ground from torrential rains.

Soils of the Ukhrul region are generally deficient in phosphorous. However, in the hill slopes, the lower portion is found to contain higher amount of phosphorous and potash, as compared to the uppermost portion.

The effect of land management measures on watershed behaviour is well reflected in run - off and soil - loss values. Hydrological behaviour of the watersheds in the above mentioned four farming system under study is reflected in the table (a).

TABLE (a)

Run - off and soil Loss From Various Farming System (Average of 1976 and 1977)

Particulars	W 1	W 2	W 3	W 4
Run - off (mm)	114.0	81.4	32.8	256.3
Run - off (p.c.)	5.3	3.9	1.5	12.9
Soil loss (t/ha)	40.9	5.8	5.0	39.9

W 1 = Jhumming ; W 2 = Partial Bench Terrace ; W3= Complete Bench Terrace ; W 4 = Puertorican Type of Terracing.

It would be seen from the above table (a) that peak floors have considerably gone down in W 2 and W 3; i.e. partially terraced and completely terraced plots. Considering the over all behavior of the watershed, W 2 appears to be as good as W 3. However, on steep hills 1/3rd of the most of the watershed may be kept under forestry.

In another experiment on topo - sequence of crops it has been observed that the highest yield of rice could be obtained when rice was cultivated on bench terraces in the lower portion, 25 per cent of the area. The study further reveals that even in bottom terraces, it would be desirable to grow rice in the lower terraces where water is available and to grow maize in the upper terraces since maize can be grown on an area with drainage.

New Technology of Agri-land-use

Lower 1/3 rd Area (Terraced) :

It is already indicated that the lower 1/3rd area should be terraced and utilized for growing field crops. It is desirable to maintain a proper topo-sequence of crops even in the 1/3rd area. For example, maize should be grown on the upper terrace, while rice can be planted in the low terrace. Then various crops can be grown in compact blocks separately.

Upper 2/3rd area :

As stated earlier, the upper 2/3rd area, in case of hills with moderate slopes may be utilized for horti - pastoral or such other farming. However, in case of steep hills it would be desirable to utilize only the mid one-third area for such farming leaving the top portion for forests, including fodder and fuel trees.

Conclusion

From the above studies, it can be concluded that partial terracing of land use pattern recommended for such hill - slopes is shown in Table (b).

Some of the advantages of the partial terracing are given below : (a) productivity can be adequately increased in the 1/3rd terraced area and improved technology can be introduced (b) soil and fertility loss can be controlled. (c) Subsidiary sources of income can be assured (d) 1/3rd terracing can be done by the family labourers of the jhummias.

TABLE (b)

Land - use Pattern in Lieu of Jhumming for the State

Slope	Portion of total area (approx.)	Land - use type	Conservation measures
Lower portion	1/3rd	Agriculture	Bench terracing
Mid portion	1/3rd	Horti -pastoral	Crescent terracing for horticultural plants
Top portion	1/3rd	Forestry	NIL

Since horticultural crops will be grown, the peasants will have a long term interest in the land - and cultivation throughout the year will bring good yield from the lower terraces. Thus permanent settlement will be induced. This system will help in preserving the ecology of the region.

Although an alternative land - use has been suggested as an immediate measure, yet it has to be kept in mind that it will be necessary to develop suitable agro - forestry - based farming system in the hill districts of Manipur. There is need to work out suitable combinations of various land - use for optimum utilization of land, water and natural resources so as to develop each area according to potentiality maintained and ecological balance and to ensure sustained products.

Appropriate Land-use Policy in Manipur

The study of land - use outlines the manifold uses of the land resources by man. Man experiments himself in the changing environment. He has been using the land from one primary use to another general use. He cleared the forest for shifting cultivation and then used the land for large - scale farming, small - scale farming, intensive farming, mixed farming, dry - farming, etc. He has used the land for one crop or a large number of crops. In this process man uses, misuses and overuses the land. "Trial and error method leaves its trace of success and failure". Ever increasing population requires revision of land utilization plans.

There are at least six vital needs of man which involve the use of land for their satisfaction. We may list them as food, shelter, work recreation, movement and security. Whenever population is dense, these become competing demands. In the context of these needs, and analysis of the existing land - use of Manipur with regard to physical, climatic, biotic and economic environs will be, it is hoped, highly useful for the purpose of future land - use planning. It is thought such studies that the region may be understood more clearly and measures may be suggested to turn it into a more productive one.

In the studies of land - use the idea of land conservation is taken as the outstanding concept. Land conservation means "using land to the optimum extent so as not only to get the optimum benefit of the resources but also to preserve it as much as possible in an unhampered efficiency for the posterity."

It is found that the Statistical data and past records of land - use in the State are unsatisfactory. No comprehensive land - use survey of the State has so far been undertaken and the figures on survey are required to be collected from a variety of sources, all of which are not compatible. Here, the author is trying to establish a systematic correlation between the past and the present conditions of land resources and to predict the land potentials for future uses . In a State like Manipur whose economy is chiefly agricultural, the utility or reliable, comparable, comprehensive and consistent statistics for the purpose of perspective land - use planning and development need not be reiterated.

Manipur State lies in the offshoot of the En Himalayas stretching between 93.2^{0}E - 94.47^{0}E and 23.5^0 N - 25.41^{0}N longitudes and latitudes respectively. The State is bordered with about 854 kms boundary line of which 353 kms are international border with Burma on the East. Physiographically the State may broadly be grouped into Valley and Hill terrain. Valley comprises 2,640 km^2 only representing 12% of the total geographical area of 22,356 km^2. The remaining 88% i.e. 19,716 km^2 belongs to the regime of hilly terrain. Based on the basis of the different physiographic spectrum the land-use pattern may be analysed as (a) Valley land use pattern and (b) Land use pattern of the hilly terrain region.

Valley land - use Pattern

As stated, the total geographical area of the Manipur Valley is 2.67 lakh hectares. About 34.0 per cent of the surface area is occupied by foot - hills lying between 1,100 metres to 1,300 metres amSL. Out of this more than 80 per cent is exposed to heavy soil erosion and soil wash. Moreover 16.5 per cent of the surface area of the valley is covered by lakes, swamps, rivers and deserted river courses not suitable for cultivation. Isolated hillocks of relatively low relief with idle, barren surfaces, between 900 m to 1500 amSl, account for 4.2 per cent of the Valley. Forests cover only 0.3 per cent of the valley. Thus, effective agricultural land - use in the valley is confined to only 36.6 per cent, while 9.0 per cent is given to non - agricultural uses.

Although the total geographical area is 2.67 lakh hectares, statistics for land - use are available for 1.52 lakh hectares or 57.43 per cent of the total. The arable land constitutes about 1,13,902 hectares or 75.0 per cent of the reported area. Mcre than 85 per cent of this is given to crop cultivation, including 4,376.46 hectares used for more than one crop. Thus it is clear that 97,447.00 hectares or 95.7 per cent of the gross cropped area is under single crop. The table given below shows an outline of the existing land - use pattern of the valley.

Existing Land-use Pattern of the Valley

Land - use classes	Area in Hectares	Percentage of the Reported Area
A. Forest Land	826.98	0.5
B. Permanent pastures and other grazing Land	3,587.56	2.4
C. Land Under Miscellaneous Tree - crops and groves and not included in the net area shown	7,255.10	4.8
D. Land put to non-agricultural uses	23,881.83	15.8
E. Barren and Uncultivable Waste Land	2,150.57	1.4
F. Land Available for cultivation	1,13,902.38	75.1
Total Reported Area	1,51,604.42	100.00

The Uncultivated Land

The uncultivated land in the table above stands for the land put to non - agricultural uses, miscellaneous tree groves, permanent pastures, forests, etc. The first category herein, i.e. the land put to non - agricultural uses stand for all land occupied by residences, community centres, roads, airports, commercial establishments, graveyards etc. These cover 15.8 percent of the reported area. It has increased from 3,701 hectares in 1952 to 21,605.8 hectares in 1976, further to 23,881.83 hectares in 1977 and still increasing the hectares. This increase is directly related to the growth of population and consequent utilization of cultivable land for residential, industrial, transport, institutional and commercial purposes. The rate of conversion of agricultural land for non - agricultural uses has accelerated by the construction of roads and service centres. Such a change in the landuse pattern of the valley, in the recent years is visible all along the National Highways, State Highways and near the service centres while the highways are constructed and repaired with the earth taken from the fertile arable lands, the towns and service centres, have been expanding haphazardly at the cost of valuable crop - fields.

The area under miscellaneous tree - crops and groves, not included in the net area shown constitutes only 4.8 percent of the reported area. This area has also decreased from 22.2 per cent in 1952 to its present meagre size. The land put to permanent pastures and grazing ground and forest land, represents only 2.4 percent and 0.5 percent respectively of the total reported area. It is to be noted that the area under pastures and other grazing lands are increasing from 2.0 percent in 1971 to 2.4 percent in 1977. But the area under forest has decreased to 0.5 percent in 1977 from 5.7 percent in 1952. This decrease in forest land is mainly caused by the continuous deforestation of the foothills.

About 4.0 per cent of the uncultivated land is covered by isolated hillocks which cannot be brought under crop cultivation, nor even under any kind of plantation. The continuous population pressure on land has a direct influence on such barren land. As a result, certain hillocks like Chingnga and Iroisemba located in and around Imphal town, have been used to install water reservoirs. It is found that such hillocks near the towns are used for the fulfilment of the urban amenities, while those near the lakes like Waithou and Loktak are used for recreation purposes with tourist homes and lodges. However, certain hillocks of the eastern border of the valley are used for plantation of pine - apple etc. Thus, the area over the hillocks left barren, has decreased from 11,139 hectares in 1952 to 2,150.57 hectares in 1977.

The Cultivated Land

The cultivation in the valley is confined mainly to the central plain area. The cultivated area covers more than 85 per cent of the available arable land of the valley. The percentage of the net sown area, as also gross cropped area, has been found to be steadily rising.

About 85.9 percent of the cropped area is under Kharif crops which include paddy, maize, chilli, sugarcane, some pulses, other fruit crops etc. It is apparent that about 99 percent of the total Kharif area is under paddy, while remaining 1.0 percent only is given to other crops.

The variation in the area under cultivation is a function of population pressure and technological achievement. It is found that at the end of the nineteenth century, the cultivated area was only 42,853.39 hectares or 16.2 percent of the total area of the valley. It then had a population of about 1,35,288 with a per capita cultivated land of 0.316 hectares. As the population growth increases the per capita cultivated land has gone down to 0.13 hectares in the year 1971. The present proportion of cultivated area of the valley is lower than the national figure of 40 percent. This decrease was caused mainly by the increasing use of cultivable land for non - agricultural purposes as stated earlier. Over and above, some of the cultivable areas, suffered from droughts caused by the abnormal monsoon conditions and lack of assured water facilities.

Land-use Policy

In the Valley :

An in - depth study of the existing land - use pattern and its problems in the valley, through analytical approaches, leads us to arrive at the following conclusions. Based on this work, the following suggestions, and hints to further works may also be put forward.

With an enormous pressure of population on the agricultural resource as expressed through an alarmingly high cultivator - cultivated area ratio and with the farthest limits reached by the extent of cultivation, further expansion of the area under plough appears limited. It would seem imperative that the cultivators of the valley should raise more than the existing quantity of harvest from the same field to meet the need of the increasing population. Therefore, serious efforts should be made for intensive use of whatever land is available in the valley. The arable land in the valley has been increasingly used for growing food grains, mainly paddy, in response to the increasing food requirement, indicating little commercialisation of crops. Unless the yield per - unit area of paddy is increased, its hectareage cannot be reduced to make room for growing other cash - crops which would derive higher income for the cultivators. Measures suggested to meet the exigency are transformation of the lingering traditional farm technology, supported by the essential infrastructure.

Intensification of land - use, particularly that in agriculture, should be achieved not due to population pressure, but by a choice of scientific farming. The analysis of crop - combination reveals the combination of a fewer crops, indicating greater productivity. During the years 1972 - 77 there was an insignificant rate of increase in the agricultural efficiency in the valley. Efforts should, therefore, be made to increase both yield and agricultural efficiency by crop - rotation and also by the use of modern inputs according to the suitability of each crop - combination region.

The crop yield per hectare may be increased greatly by human efforts in a variety of ways, such as, intensification of cropping, application of bio - chemical fertilizers and manures, use of improve and HYV seeds and irrigation water, adoption of corporation and crop - diversification techniques, and control of pests and diseases. This will not result only in increased productivity but also change the character of land - use and crop distribution pattern of the valley.

Soil conservation and soil survey department should take various measures such as afforestation of the foot - hill and catchment areas of the Manipur Valley. The government should be fully alive to the problem of soil erosion and take appropriate steps whenever necessary. Detailed soil survey and application of adequate modern bio -chemical inputs accordingly, are essential. It is, therefore, suggested that soil testing centres with modern equipments be set up in different physiographic situations of the valley to ascertain the properties of the soils, so that necessary measures can be taken for their development.

Besides, soil erosion and soil waste, siltation, flash flood and drought are the common natural hazards hampering an optimum land - use of the valley. Rabi cropping is made difficult during the prolonged winter drought. Failure of monsoon plays havoc with the kharif crops which are vital to the life and economy of the valley. Irrigation, in order to combat the adverse effects of both flash flood and drought, becomes necessary.

Little attention to soil conservation has resulted in a great misuse of land in some parts of the valley. The land - use pattern needs reorientation with a feasible and dynamic plan for the optimum use of every hectare of land. The paucity of land under forest is now realised. It is, therefore, suggested that the area under forest be increased at least to 33 per cent of the total area of the valley, particularly in the foot - hill and catchment regions and on the isolated hillocks, by afforestation.

There is a great need for development of pastures, which are, as such, totally absent, except as a few patches of degenerated scrub - lands and marshes. If the pastures are developed there is a prospect for developing dairy - farming in the valley.

As the protection of crops from pests and diseases is beyond the capacity of the individual cultivators, the government, should take measures for appropriate policy on adoption of natural farming on bio - fertilizer at the appropriate time. It is also necessary to provide essential apparatus and eco-friendly chemicals for this purpose, at subsidized rate.

Industrialisation and development of tertiary sectors are so insignificant in the State that they have not been able to divert the surplus man - power from agriculture to other occupations, Therefore, urgent steps should be taken to develop the secondary and tertiary sectors of industry in order to make the agricultural land - use of the valley economically more efficient by withdrawing the excess man - power from it.

Agriculture is the mainstay of the economy of the valley. The malady - ridden traditional socio - economic structure of the peasants must be rationalized and reorganised into a new pattern to fit into the productivity oriented economy of the modern world. Earnest efforts should be made for providing modern agricultural infra - structure, which includes among others, rural electrification, agricultural credit, transport and communication, etc., for optimum land - use.

The rate of literacy among the agricultural population of the valley is low. It is one of the most important factors of agricultural efficiency. Therefore, the percentage of literacy in the rural area must be increased by strengthening the adult education, farmer's functional literacy programmes and effective implementation of universalization of elementary education.

As the available productive land per capita in the valley is bound to decrease with the increase in population under the existing law of inheritance, the latter should be amended to prevent further decrease in farm size. Besides, new laws should be framed and introduced to debar the non - cultivating adult members of a family employed in other occupations from inheriting and purchasing agricultural land.

The Rayatwari system of land tenure and various land reform measures of the government have indirectly led to the creation of small peasants by increasing the number of small holders. The inadequacy of land , especially the smallness of holdings, has created a serious problem in the land-use development of the valley. There is, therefore, only one alternative to eliminate this contradiction of the present government policy and that is the creation of joint co-operative farming under the land consolidation programme and pooling the means of production of the small individual cultivators together in order to make farming an economic proposition. Co - operative farming with the State patronage is more suitable for an underdeveloped region like Manipur valley, where majority of the farms are of economically non - viable size.

Unless the farm size is raised to an economically feasible unit, adoption of improved technology is not possible. Therefore, efforts should be made to consolidate the fragmented land holdings. For consolidation programme the following steps in stages may be suggested, (a) fragmentation of holding should be prohibited by legislation, (b) the name of all the genuine cultivators should be registered, (c) non - cultivators should not be allowed to possess cultivable land, (d) agricultural land of the State may be nationalized, (e) lands of economic size may be redistributed among the registered cultivators.

The land records and settlement reports of the area are not Systematic. Therefore, it is very essential to compile and maintain land records properly. Without this, no proper land - use and land development planning will be fruitful. A land - use survey and planning wing of the government may be useful for economic development of the State.

The analysis reveals that the existing land - use pattern of the valley is the product of unplanned and unorganised use of land. The deep - rooted malady in the land - use can be cured through an interdisciplinary and integrated approach taking the land records and land - use data as the base materials. The responsibility of an overall land - use development and the implementation of the remedies cited above, lie partly on the policy makers and partly on the peasants. The leaders and the people at large should be sensitive to these acute problems faced by the land - use development programmes of the State and there should be no scope for complacency or personal bias.

Soil Erosion & Environmental Problems

Soil

Some 350 million years ago, plants and animals previously confined to the sea began to establish themselves on the land surface. They had difficulty at first in adapting on the barren rock surface. This was a period of survival of the fittest; no species of plant or animal could survive if it did not adapt to the new environment. Although, the plant and animal tried to survive on the barren rock, they died soon. Their remains added to the rock body. And through millions of years with this similar process a thin mantle of soil was accumulated over the barren rock. The first recorded information about the soil dates back to ancient Greece when Aristotle and later his disciple - Theophratus (IV century B.C.) made some interesting observation about the properties of soils as related to plant nutrition.

However, one of the first often subconscious acquaintance a person has with nature is the soil. But there are few who can state what SOIL is. A soil is made up initially of rock that has been substantially transformed by the vegetation and various micro and macro - organisms.

The influence of the organism have on rock depends on the climatic conditions, especially on the hydrothermal regime. The process of soil formation not only depends on the rock but also on the plant and animal organisms, the climate, and the age of the land - mass. Hence, we may define it to be a natural formation resulting from the transformation of surface rock by the combination of the climate, plant and animal life and ageing. In a sense soil is forming in a cyclic way. And the cycle is based on the continuing exchange of matter and energy between living organisms and their products, and the rock beneath them. This exchange results in the formation of a soil. The cycle can be described thus, plants and animals accumulate solar energy, and when they die, their remains decay releasing energy. On the one hand, new mineral and organic compounds come into being due to the decays, while on the other hand, the primary minerals comprising the soil forming rocks are transformed. As a result, heat is released. These processes of synthesis and decay can be quite diverse and depend on the conditions under which soil is formed.

The result of the continuous exchange of matter and energy between living organism and rocks, this exchange being mediated by the atmosphere and hydrosphere, is the formation and evolution of soil on the earth's surface.

Soil & Human Being

When primitive man came into the picture, he, like other animals, was forced to live in a balanced manner that nature demanded for survival. As man becomes the so called civilized, his demands upon the land for additional food and clothing grew. He changes from nomadic life to cultivating the soil and leads to the destruction of natural ecosystem. Therefore soils are more exposed to the erosive forces. In this context we may look back to the lost of civilizations like Mesopotamia, Syria, Lebanon, Egypt, Indus Valley, etc. etc.

From the lost of the above mentioned civilizations, it is clearly evidenced that reckless use of land and excessive deforestation leads to the devastation of soil and civilization.

Problems in Manipur

The State of Manipur is experiencing the ecosystem degradation of the most severe order which is affecting the hill tops & slopes, water channels, Manipur valley and the wet lands. It is also a fact that most of the cultivable hill slopes were seriously affected by the problems of soil erosion and soil wash. The key factor of all these hills lies in the problems of land - use patterns in the Hills of Manipur. Further these problems have been aggravated by the expansion of shifting cultivation (Pam Lou) in the hill tops and steep hill slopes.

Consequently the forest coverage has shrunk and the soil erosion of a most severe type has been initiated on the hill tops and hill slopes which is choking up the water channels and also silting up the wet lands in the Manipur valley. On the other hand, the continuous growth of population makes normal Jhum cycle tremendously reduce to 2/3 years. The shorten Jhum cycle has caused extensive damage to the natural vegetation in the hilly terrain of the State. This practice of cultivation has completely destroyed the forest coverage and the soil has been seriously impoverished and in consequence many hill areas of Manipur have been out - cropped and they have become unsuitable for bearing any vegetation. Thus shifting cultivation affects the entire economy of Manipur State as a whole. And also is creating a serious ecological degradation in the State.

Land & Forest Management

Sustainable development always link with the kind of growth and development which co - relates with the policies of natural environmental factors. In other words, developmental activities integrated with the environmental policies is one of the vital points of sustainable development. Sustainable development and its technical know - how is essentially the study of relationships regulating interactions and the schemes with regard to growth. It has become increasingly apparent that the development models establish a mathematical relationship between living beings and abstract entities. The analysis of development must be intelligible to locate the relations between necessities and freedom. The gainful utilization of the resources for human welfare are well delineated and the need for controlling population explosion, resources shortage and environmental imbalances leading to natural calamities is stressed upon.

On the whole, basically there are three components of sustainable development- the preservation of environment, equitable development and family planning.

Since the dawn of civilization, food, fodder and fuel needs have been met by the constant use of land. But as the days gone by, the pressure of population demands more from our available natural resourçes. As a result, our lands are being extensively depleted of its green coverage and exposed to degradation processes thereby resulting to loss of productivity to the vast area of our available land resources.

The problem of environmental degradation and its adverse impact is specially acute in the North Eastern Region. This is the compact Zone of the Eastern offshoot of the Himalayas marking for the highest record of heaviest rainfall region in the world. On the contrary the region is insufficient for safe drinking water for the inhabitants of this region. Most of the agriculturists are suffering from lack of sufficient agricultural water and also of excess of water. Every year this region is facing problems of flood and drought alternately during the same year. The region is mingled with the series of serious problems of soil erosion, landslide, siltation in the lakes and riverbeds. These are the direct products of environmental degradation in the region. The present day environmental degradation is the out - come of the age old practice of land - use in the region. As a result most of the forest area of the State of Manipur has been degraded upto its maximum limit. Neither the people's economy developed, nor poor people and its society could afford environmental conservation. Will they be able to invest the expenditure required for the pollution control as other eco - system implies? On the other hand, the cost of soil erosion can only be measured in terms of the resultant loss of agricultural produce etc. Then we may also realise the cost of agricultural practices which clean air, regulates water - flow, fixes nitrogen, absorbs pollutions, produces biomass and provide habitat to other flora and fauna.

Deforestation results in rapid soil erosion which has taken the forest thousands or millions of years to build its natural eco - system. It also disturbs the hydrology of the area/region and causes climatic changes. These are the only scenario faced by all of us for every day in our hills and plains of the State. The problems of deforestation continues even if we all know that it is detrimental to our environment. What is the main reason for deforestation in the State? It is the economic condition of the people of the region. Since the evolution of human settlement, people of this region have been living in the forest and earning their livelihood from the forest only. Till now majority of the people depend on forest for their livelihood by exploiting the timber and cultivating the forest land. This may be classified into two major activities i.e. (1) Jhumming and (ii) Logging. Logging is increasingly a recent phenomenon and it is being done indiscriminately. The increasing tendency of Jhumming and shortening of Jhum cycle to about 2/3 years is the direct impact of the increasing population pressure on the limited land areas. It is also recorded that more than 950 sq. km. of forest land have been deforested annually by the Jummias. On the other hand, the logging business has also been increasing enormously. By combination of these two activities the total dense forest area of the State has been reduced to only approximately 3.2% (DSTE : 1990). To maintain a pleasant climate with timely rainfall and also to enrich the soil fertility and supply of safe drinking water in an area, it needs at least 33% of the total geographical area of the State to be preserved as a forest coverage zone (UN).

The real wealth of the people lies in its natural resources. Our people know well about the value of each and every species of the flora and fauna of their land and forest. Therefore, the only real solution of the above mentioned environmental degradation problem must be generated only from among the people of the State. No imported technology and design will be able to solve the problem of ours permanently. In this context, we may call for proper management of available land and forest with the idea of conserving bio-diversity and improving the environment.

Action to be taken through People's Involvement

1. Land - use survey & resource Mapping
2. Land potential classification

3. Watershed Management planning
4. Watershed on catchment area as a planning zone
5. Waste land & wet - land Development
6. Introduction of Natural farming in lieu of Jhumming practices.

References

1. Sharma, T.C., 1976, "The pre - historic background of shifting cultivation", proceedings of Seminar on Shifting Cultivation in North East India, p.p.1-4.
2. ICAR, 1981, "Shifting cultivation in Manipur and Nagaland" proceeding of the Agro - Forestry Seminar, p.132.
3. Saha, N. 1973, " The Economics of shifting cultivation in North East India", Adhoc study No.22 Agro - economic research centre for NE India, A.A.U., Jorhat.

   1976, " Levels of Production and Income under shifting cultivation in NE India". Proceeding of Seminar on Shifting cultivation in NE India, pp.15 - 20.
4. Schlippe, P.D. 1956, Shifting cultivation in Africa, Routledge & Kegon Paul, London.
5. Majumdar, P.S. 1971, Plan programmes of hill development in North Eastern India, Indian Farming XXI (v), pp.5-9.

5

Managing Environmental Resources in the 21st Century

A study on environmental planning of this part of the country has to be partly theoretical and partly empirical with greater emphasis on the latter. There are several theories regarding the planning and forecasting of future based on the present existing indices. This is a geographical approach in the sense that the problems are viewed from the spatial angle and from point of land - use and linkage of the growth of population in the State to the arial extent of the arable or cultivable or forest land and the spatial distribution of population within the state are the main theme of this paper. The paper also focuses attention on the physiographical factors which influence the internal structure and external projection of the land - use pattern of the state.

The present paper is planning oriented and gives guidelines on the planning designed to meet the demand of ever increasing population at present as well as in the 21st century while maintaining a sound eco - system in the State. It also provides necessary guidelines for planners to do the comprehensive socio - physical and environmental planning of the State.

Empirical analysis of this type attempts on had been made in the past by several scholars of different disciplines focusing attention on different aspects of planning. Some of the geographers have, however, tried to integrate the concepts of development in different disciplines and to suggest a theoretical framework which provides solutions to the land - use problems - both physio - social and environmental.

Planning for the State has to be both clinical and preventive. As a clinical endeavour, it is designed to solve the existing problems. It would look at the region from various angles and suggest, after proper and systematic diagnosis, the steps to be taken to put the system in order.

To an amateur planner the two aspects of planning may appear to involve separate policies and hence needing different approaches. This, however, is not true. Both for clinical and preventive purposes the planner has to examine the problems that the State faces, and has to understand the forces - Geographic, Economic, social ; technology and others which have to observe the forces that are likely to work in favour of or against the development of the State in the 21st century. He has to understand the institutions and the limitations within which he himself has to function and also know the nature of services the community needs today and will need in future.

Most of the development planning organisations deal only with physical aspects of planning. This is especially so in India. But we know that physical planning can succeed only if it is based on realistic assessment of the socio - economic problems and issues involved in the land - use planning of the State. A planner must get sufficient data on the present and the future needs of the State before he can prepare a good physical plan. He must know the present and future problems associated with housing, work, movement, recreation and education, before he can even think of a physical plan. Above all he must know the socio - economic needs of the State before he can think of an effective and meaningful use of land.

To get the data needed for the plans, all planning endeavours have to be backed by realistic diagnostic surveys by specialists in various fields. Absence of such studies makes a physical plan nothing more than an attempt to solve the problems before knowing what the problems etc are. If planning efforts have so far failed to raise the economy of the State, it may be because the efforts were not based on sound diagnosis of the problems. Hence, an Alternative Planning Approach has been designed through the following process.

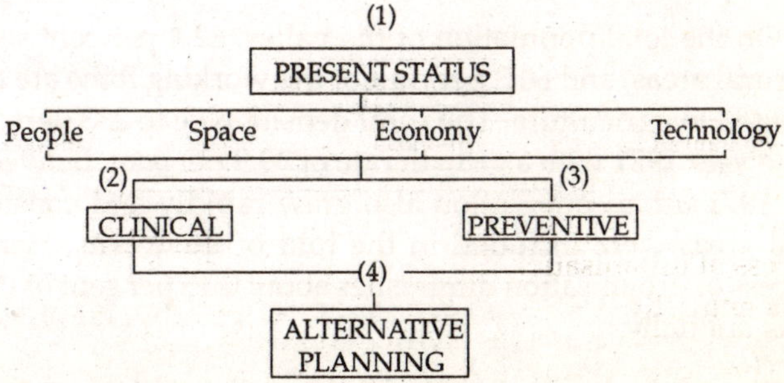

Valley Environmental Resources & People

The total geographical area of the valley is only 2,630.8 km^2 or 12 per cent of the State's total area (22.347 km^2). It contains 69 per cent of the State's population (1991). The very low density (17 persons/km^2) of population in the hills of Manipur is mainly due to rugged topography, made up of steep slopes, forests and consequent lack of agricultural opportunities, mineral and other resources having remained unexlored so far. The lack of suitable transport system has added to the adverse economic condition of the hills.

According to 1981 census, the arithmetic density of the valley is 300 persons/km^2. which is much higher than the all India average of 167. However, this figure does not give the exact picture of human density in the valley. Because, out of the total valley area about 546 km^2 is occupied by lakes, swamps, marshes, barren uplands and hillocks. Eliminating the negative area covered by these features, the density of the valley is calculated at 400 persons/km^2. The available arable land of the valley is only 961 km^2 or 36.4 per cent of the valley as a whole with an agricultural density of about 145 persons/km^2 (1981).

The arable area of the valley is only 96,046.153 hectare which accounts for only 0.13 hectare per head. Hence eight persons have to depend on a hectare. The present problem is further aggravated by the high growth rate (3.4 percent per annum) of the population of the valley. It is estimated that the projected population of he valley in 2001 A.D. will be approximately 2.08 millions of people. Thus it is expected that with the population raising at the present rate, the pressure on land will be tremendous.

On the total population of the valley, 62.4 per cent settled in the rural areas, and 60.3 per cent of the working force are directly engaged in agriculture. The rural density rose to 233 persons km^2 in the year 1971 with a growth rate of 39.9 per cent. Between 1951 and 1971 urban population also grew rapidly and considerable rural areas were included in the fold of the towns. Hence, the process of urbanisation diminishes about 0.15 per cent of the rural areas annually.

The agricultural density of the valley also rose from 132 persons/km^2 in 1961 to 144 persons/km^2 in 1971. The economic division of the population constitutes 29.9 per cent. The non working population accounts for more than two thirds putting an immensely heavy burden on the former. This section includes the population belonging to the age groups of 0 to 14 years and 60 years and above. Moreover about 80 per cent of the State's total student population are concentrated in the valley itself. Above all more than 99,671 persons under the age group of 15-24 years have been enumerated as dependents. They include educated unemployed and those undergoing higher studies.

Space Problem

Various limitations and impediment in the utilisation of the available land resources of the valley are inherent in its physiography, soil conditions, climatic characteristics and socio - cultural milieu. Besides, the increasing pressure of population on the agricultural land, excessive sub - division and fragmentation of the holdings have reduced the per capita available land for agriculture in the valley to very small parcels. This situation has reflected itself in very low intensity in respect of agricultural land-use development. On the other hand, population growth on the valley is creating a serious problem of unemployment among the peasants.

The increasing population needing more and more land for both agricultural and non - agricultural uses has posed a serious problem in the land - use planning of the valley. There are two ways of which agriculture output can be raised, firstly, by extending the area under cultivation and secondly, by raising the output per hectare.

The prospect of further increasing the net sown area is extremely limited, for almost all the arable land of the valley has already been brought under the plough. The areas under both forest and permanent pasture are decreasing fast. It has been mentioned that the per capita cultivated land decreased from 0.325 hectare in 1951 to 0.125 hectare in 1971. It had further decreased to 0.08 hectare in the year 1976. By now the situation has been further aggravated.

It has also been mentioned that a higher increase of population (39.92) is recorded during the decades 1961 - 71 and 1971 - 81, while it was only 19.9 per cent during 1941 - 51. Based on the FAO's recommendation of 0.5 hectares of cultivable land per person the present population of the valley requires extra arable land over and above the existing arable area of 72,306 hectares (1981). If it is calculated for the projected population of 21st century the required extra arable land will be 9,41,998 hectares. It is obvious that no expansion of arable land is feasible. In fact, the scarcity of land has been further aggravated by the sprawling of towns and associated infrastructural developments which have been eating up 0.15 per cent of the arable land annually. However the only possible way of increasing the area under plough to some extent is to reclaim the culturable waste, utilize the area left fallow and practise double and multiple cropping. The cost of reclaiming new land is hardly justifiable socio - environmentally, however economically desirable may it be. Nor does the expansion of the cultivated hectarage, as such, contributes to agricultural productivity. Thus, new plans for agricultural development should concentrate on intensifying the output within the existing arable area of the valley.

It is found that more than 98.5 percent of the total cultivated area of the valley depends on rain water supply. Most of the rain water during the monsoon goes away as run - off and is not preserved. Only a little portion of the rain water could be used by the peasants in the main paddy season. During the drought as also in the dry seasons, most of the area under plough is seriously affected hindering double cropping. In fact only 2.3 percent of the total cultivated area of the valley was under more than one crop in the early eighties. It is thus obvious that lack of assured water supply pose an acute problem on the agricultural land - use.

Nearly 99 percent of the peasants of the valley work on their land with a pair of bullocks and a wooden plough. Thus agriculture in the valley is carried out at a level even below the bed rock of the present day mechanization of agriculture. Traditional implements and animal power are still widely used in the valley. Poverty and ignorance coupled with poor extension services have served to keep modern agricultural implements out of the farmer's reach. The peasant has often to borrow or hire even the implements which he knows how to use. Oxen and buffaloes have still remain as the main Means of plough. Thus it is essential to preserve adequate grazing land. But acute population pressure accounts the low hectarage devoted to fodder crops. Nor do the livestock have any satisfactory alternative source for proper feeding. Thus so long as animal remain the source of plough power, the progress of double cropping in the valley will remain insignificant.

Hilly Terrain Environment : An Alternative Planning Approach

Plan Framework :

Spatial considerations : The basic need for preparation of any plan is a spatial address. This can be identified in two ways. While one recognises administrative jurisdictions the other takes cognizance of topographical characteristics leading to watersheds, as planning units or regions. The former has an advantage of ease in administering the project because the planning unit coincides with a prevalent administrative unit. However as the ecological system does not conform to administrative boundaries this approach has an inherited limitation. This drawback may effectively be handled by adopting the watershed approach. Also the administrative boundaries being comparatively more controllable than the ecological phenomena it is better to priorities ecological phenomena, considerations over administrative ones for selection of planning units. Once it is agreed to adopt a watershed approach, the mechanism for delineating the units needs to be worked out. It is proposed here to demarcate the whole of the hilly region of the State as catchment area of two major river systems i.e. the Barak and the Chindwin system, both of them, implying as a two separate planning regions.

While the scope of the above discussion is valid over different ecosystems, the watershed of a particular tributary of the above mentioned systems may be defined as a planning unit in our context.

6

Global Environmental Concerns

Introduction

Environment comprises the whole range of external influences acting on organism in both the physical and biological forces of nature surrounding the human beings. All living beings form an integral part of a balanced and intricate eco-system. They draw their sustenance from the solid, liquid and atmospheric resources of mother earth. All of them undergo the natural phases of reproduction, sustenance and extinction. However, the ultimate source of all kinds of energy for the entire eco-system is the SOLAR ENERGY.

The eco-system consists of a network of sub-systems such as seas, forests, water shed/catchment areas and wetlands etc. Plants, trees, insects, animals and all other organisms in a forest are interconnected in a network of devourer-devoured relationship. None of these elements and organisms in the network is useless or purposeless - each of them helps to reegulate the balance of the sub-system. For example, bio-gas-chemical cycles are essential for the process of reproduction in the physical environment. Another illustration can be that of the natural cycles of water, carbon, oxygen are fused into the environment, absorbed and used in biological constituents and again emitted into the environment.

All the constituents of the biological system on earth are reproducible. The process of reproduction is held in balance as long as the natural cycles, their interrelations and the hierarchy of the food-chain are not disturbed. The balance of the eco-system is on the earth, such as stones and minerals etc., are not reproducible and their stock is limited. Hence their use has to be very carefully regulated and monitored keeping the future in mind.

In all, environment has been correctly defined as "the aggregate of all the external conditions and influences affecting the life and development of an organism." Human beings form an integral component of the ecological system of the EARTH PLANET. It is necessary, here, to distinguish the two related terms : environment and ecology. Generally, 'environment' refers to the surrounding in which we live in, whereas 'ecology' is the relationship between the various factors that make up our environment. The earth's biosphere is maintained by a delicate balance of these factors. Any change or breakdown in this system of interdependence can lead to a chain reaction, resulting in catastrophic phenomena and may even endanger the very existence of human kind.

Until recently, the planet Earth was a large world in which human activities and their effects were neatly compartmentalized within nations, within sectors like energy, agriculture, trade, and within broad areas of concern like environment, economic, social etc. Hence, at this present juncture one should have to look to the problem as one i.e. Earth as a planet. Again we may realise the fact that the Earth is one but the world is not. We all depend on the biosphere for sustaining our lives. Yet each community, each country-state, strives for survival and prosperity with little regard for its impact on others. Some consume the Earth's natural resources at such a rate that would leave little for future generations. Others, many more in number, consume far too little and live with hunger and disease etc. SImultaneously, stress on natural environment has been the result of the growing demand on scarce resources and the pollution generated by the rising living standards of the relatively affluent. Poverty itself pollutes the environment. Those who are below the poverty line will often destroy their immediate environment in order to survive. They will cut down forests; they will overuse marginal land; and growing in numbers, they will crowd into congested cities as slum dwellers. Thus today's environmental challenges arise both from the lack of forms of economic growth.

Further, to study the global environmental concerns, we first need to understand the conditions of water resources of our earth. In this context, it would be more appropriate for us to have thorough look into the water resources at this instance.

How much water we have

As viewed from space, Earth appears as a watery planet because nearly 71 per cent of its surface is covered by oceans. All the water on Earth makes up what is called the hydrosphere. The origin of the hydrosphere dated back to the beginning of Earth's history more than 4.5 billion years ago. At some early stage of the Earth's formation, it began to release water and various gases from it's interior by a process known as degasing. Most of the water on Earth's surface is concentrated in the southern hemisphere where the water to land ratio is about 4 to 1 in contrast, and the distribution of water differs from place to place and from region to region.

Now-a-days, the problem of water is becoming increasingly acute mostly due to the unscientific exploration of available resources and also destruction of the natural environment. The problem is further aggravated due to the deterioration of quality of surface waters and their progressive pollution. However, a good deal of effort is being made to prevent this evil but its efficacy is so far inadequate. It is also realised that the basic cause of the environmental degradation and problem of water is mainly caused by the consequent effect of the continous population growth and expansion of urbanization and its infrastructure activities at the cost of the Natural physiographic regions.

The many proposals for rational use of water resources and its conservation may be categorised into two basic groups. The first includes, in the main, the study of universal principles of water cycle and global water budget. The approach in the second group proceeds from the assessment of the water resources and nature of utilization etc. For the rational use of the water resources and its conservation a fundamentally different approach is necessary for organising the use of water resources and measures to protect them. The most important approach is the end of the abuse of water resources and proper taping of available resources for the whole mankind. To substantiate this concept we shall try to prove, first that the water resources available on earth, provided they are used prudently can meet the requirements of mankind and, secondly, we shall outline the principal ways to safeguard and use them rationally.

The World's Water Resources

The water resources consist of statistic resources of the hydrosphere and water continually reproduced in the process of their circulation. The water is found in the atmosphere on the earth's surface and within the crust of the earth. In the atmosphere, water is found in the wormiest layer, the troposphere, either in vapours or solid state, or in the form of liquid droplets. Surface water; i.e. the water in oceans, seas, lakes, river reservoirs and glaciers, is found either in liquid or solid state (snow, ice). Within the earth's crust i.e. within the lithosphere, water is found in vaporous, liquid or solid state; moreover, it can be firmly or loudly bound by rock particles (hygroscopic, pellicular water), as well as bound chemically in the crystal lattice of various minerals. Surface water, underground water and the water contained in the tissues of all animals and plants (BIOSPHERE) together compose the water envelop of the earth usually called the Hydrosphere.

The total volume of resources of water at any given time is not great but, owing to water circulation, they assume tremendous importance for the life of human race. They refer, primarily, to river water whose volume within river beds adds upto a mere 1,200 cu km (less than one millionth of the entire hydrosphere). However, owing to its extremely high mobility, river water renews in river beds once in 12 days on the average, or 30 times in one year.

The volume of river bed waters taken at a time could supply a little, a half of mankind's annual demand for water circulation, however, keeps water in constant motion, while its volumes involved in this circulation are continually expanded and restored. This remarkable property maintains uninterrupted restoration of water resources, which puts new masses of fresh water at mankind's disposal.

Indeed, soil moisture, which is essentially important for the biosphere, is a component of hydrosphere. The rate of its renewal is about once a year. Lake water is restored at a still slower rate, once in about ten years on the average and the renewal of ground waters occurs at approximately once in 5,000 years while deep stagnant salt water layers being practically excluded from the

Global circulation of water. The rate of water circulation in the World Ocean is about 3,000 years with almost 45,300 cu km of water evaporating from its surface annually. Indeed, owing to the water renewal in the ocean is much higher. For example, it takes about sixty years for the entire ocean water to be replaced under the action of currents. The mass of polar glaciers is replaced slowest of all. With its annual loss of 2,900 cu km to the ocean the mass of the glaciers renews roughly once in 8300 years.

The largest fresh-water storages on the earth are represented by the Glaciers, which occupy over 16 million sq. km. i.e. about 11 per cent of the total land surface. The greatest thickness of the continental ice has been recorded in Antarctica, where it averages to 2,000 m. while the maximum thickness is 4,200 m (Soviet Glaciologists). If all the ice present on the earth melted the world sea-level would rise 64 m and the area of the WORLD OCEAN would extend for an additional 1.5 million sq. km. approximately while the continental area would correspondingly be reduced by one per cent.

Some approximate figures by M.I. Lvovich (Russian Scientist) give an idea of the annual water budget of the globe as follows :

TABLE -1

Total Amount of water on the Globe and its Distribution Among the Different Natural Spheres

Portion of the hydrosphere	Water volume 1,000 km^3	Percent of total volume
World's Ocean	1370323	93.96
Underground water (including that in the zone of active water circulation)	60000 (4000)	4.12 (0.27)
Glaciers	24000	1.65
Lakes	2801	0.019
Soil moisture	852	0.006
Atmosphere vapour	14	0.001
River water	1.2	0.0001
Total	145470	100

1. Including about 5000 km^3 of water in storage reservoirs.
2. Including about 2000 km^3 of irrigation water.

TABLE 2

Annual Water Budget of the Globe

Water budget components	Annual volume km3	Annual depth mm	Percentage of the global precipitation rate
Exorthetic Continental Area (11680 thousand km^3)			
Precipitation	106000	910	20.2
River runoff	41000	350	7.8
Evaporation	65000	560	12.2
Endorheic (closed) continental area (32100 thousand km^2)			
Precipitation	7500*	238	1.4
Evaporation	7500	238	1.4
The World Ocean (3611100 thousand km^2)			
Precipitation	411600	11400	78.4
River water inflow	41000	114	7.8
Evaporation	52600	1254	86.1
The Global total (510000 thousand km^2)			
Precipitation	525000	1030	100
Evaporation	525000	1030	100

* Including 830 km3 of the river runoff.

Water Cycle and its Pollution

Atmospheric, surface and underground waters are closely interconnected and are found in continuous motion. Solar energy and gravity together cause a ceaseless moisture circulation. After being evaporated from the surface of oceans, seas, rivers, lakes, dry-land and plant & vegetable cover, water enters the atmosphere in the state of vapour carbon dioxide and sulphur dioxide, and this figure is likely to reach a staggering 50,759 million tons by 2000 A.D. Soil is being lost from the land surface at the rate of 2.5

billion tons every year due to erosion alone. A major portion of water bodies on this earth is becoming increasingly polluted as a consequence of rapid industrialisation and excessive population growth.

At one time no one questioned the amount of fresh and pure water present on this globe. After travelling for a long distance a man did not have the slightest doubt in quenching his thirst with water from a nearby lake or river. The problem of that water being loaded with toxic chemicals or being heavily contaminated with the pathogenic microbes did not arise.

Now, the whole scenario has changed completely. The problem of pollution is intensely acute in many parts of the world. Hundreds of rivers, lakes, streams and other water bodies have now become polluted with all sorts of industrial, agricultural and domestic wastes, oil and city sewages laden with numerous pathogenic microbes. A large number of lakes and water ways has become increasingly murky, foul - smelling and choked with excessive growths of algae. The world's industrial products now include approximately 30,000 chemical substances and several hundred new substances of this type are added to the list each year. Public health authorities in many countries are greatly alarmed by various toxic chemical pollutants being discharged from industries producing these chemical substances, and ultimately finding their ways to various water bodies. The pollutants include flourides, nitrites, nitrates, selenium, lead, chromium, cadmium, mercury, chlorinated hydrocarbons and various biocides. According to the IRTC, the total volume of water polluted by industrial and household wastes upto 1970 was 9,145,000 million tons and this amount will increase by 3.6% by the end of this century.

If used for drinking, polluted water may transmit various intestinal infections like typhoid, cholera and dysentry, and also jaundice and viral diseases. In 1986 during the Kumbhamela a group of scientists analysed samples of water of the River Ganga at Haridwar and found, in utter dismay that the level of certain pathogenic microbes was 1.5% more than tolerable limit. According to the Food and Drug Agency (FDA) of the USA natural waters in 33 out of 50 states in that country have become unfit for human consumption without prior special treatments. As a result of acid rain, caused by oxides of sulphur and nitrogen emanating from coal - fired industries and millions of automobiles on the road in Central Europe and Great Britain the pH of water in one fifth of nearly 50,000 lakes in Sweden has dipped to such an all time low that marine lives in them have now become extinct. Due to this very factor vast tracts of pine forests in the *Appenines* Italy have withered, and West Germany has been denuded from one third of her forest area.

As the pace of human civilization becomes faster the demand for fresh water also continues to gallop. A man daily needs on an average of 2 to 2.5 litres of fresh water just to live. If his other activities are taken into account, the daily need soars to roughly 300 litres. Above this, a large amount of water is also required for his agricultural and industrial activities. For example, a field of one hectare requires about 3 million litres of water for growing maize. In the same way, a similar field requires 8 million and 12-20 million litres of water for growing cabbage and paddy respectively.

The run-off waters from such and other modern agricultural fields contain a perceptible quantity of phosphatic fertilizers, seed sterilants like toxic mythyl mercury dicyanodiamide (MMD), and various herbicides and insecticides mainly in the form of D.D.T., B.H.C., heptachlor, eldrin and dieldrin. Such polluted water may cause damage to liver, kidney and nervous system of human body. DDT and BHC are also carcinogenic and mutagenic. In the Central American countries of Honduras, Nicaragua and El Selvador about 4000 poisoning cases from pesticides are reported every year. Thus when these chemicals finally find their ways into and pollute waters of nearby ponds, lakes and rivers they become an undeniable health hazard in various manners and stages.

Like the need in agriculture, a huge quantity of water is indispensible in industries too. An atomic power station of a million kilowatt capacity requires 3 million litres of water every minute. 18,000 litres of water are required in refining 1 ton of crude petroleum. Likewise, smelting processes to obtain one ton each of pig-iron and copper will require 200,000 and 500,000 litres of water respectively. A minimum amount of 4 million litres of water is a must in the production of 1 ton of synthetic fibre.

Under these circumstances it is becoming a herculean feat to avail of everyone on this earth with enough potable water. It is said that drinking water will become as costly as petrol in certain parts of the world by the end of this century. In some far flung areas of Rajasthan a casual visitor from outside may be welcomed with a glass of milk, but he may notice a conspicuous reluctance from his host when he asks for a glass of water. According to a W.H.O. Compilation, three out of every five people in the developing countries do not have access to safe drinking water at present. Consequently every year 250 million people are afflicted with diseases like typhoid, dysentry, cholera, viral hepatitis, trachoma, etc. resulting in the death of some 25,000 people every day. During the period from 1981 to 1990 A.D. which had been declared by the U.N.O. as the International Water and Sanitation Decade, 360 million people in India had their dream of getting safe drinking water still remained unfulfilled.

In India the pollution of various water sources has reached an alarming proportion. For instance, the River Ganga, which has been inextricably associated for centuries with the Indian culture, and with an age old belief that a single holy dip in it will wash away 108 sins from a person is now so polluted at many places along its bank that simple bathing in it may invite the wrath of various diseases. At Triveni Kuntighat, 65 kilometres upstream from Calcutta, the Hooghly River is so polluted that crops in the nearby fields started withering when its water was used to irrigate them. Twenty years ago from today, the river bank here was crowded by youths who swam and played for hours in the river. Now, it is said that if a man simply takes bath there for three consecutive days he is sure to get bed-ridden with a severe symptom of influenza. The prime culprit for all these illnesses is an industrial establishment to be described later on.

The main cause for the pollution of the Ganga are the industrial effluents and sewages discharged from a large number of industrial establishments and cities and half-burnt dead bodies thrown in the river from the numerous burning ghats dotted along its banks. Everyday Calcutta alone produces 397 tons of industrial and domestic wastes out of which 42% i.e. 167 tons are discharged into the Hooghly River, a distributary of the Ganga. Of all sorts of pollutants dumped in the Ganga, 85% emanates from 17 cities of class I grade such as Varanasi, Kanpur, Allahabad, Ghazipur,

Patna, Monghyr, Bhagalpur. One industrial establishment that can be cited among others contributing a significant share in polluting the Ganga is the Keshoram Rayons of the Birlas producing synthetic fibres, situated at the aforementioned Triveni Kuntighat. The wastes discharged by this factory in the river contain among others sulphuric acid, carbon disulphide and compounds of zinc, lead and mercury. It has led to the inability by a large variety of fish to spawn and the consequent drastic reduction in their numbers. Thirty years ago fishermen used to catch daily 5-6 quintals of hilsa fish here. Now they have to be contented with an amount as meagre as 30 to 40 kilograms, that too contaminated with zinc, lead and mercury, which may lead the consumers of those fishes being afflicted with various ailments of unknown consequences.

According to the National Environmental Engineering Research Institute (NEERI), Nagpur 70% of India's inland water is unfit for human consumption. Diseases like cholera, jaundice, dysentry, typhoid, gastroenteritis, hookworm and guinea worm infections are all transmitted to man and animal through the agency of water. After heavy rainfall during the monsoon season this year many parts of Delhi, especially the trans Yamuna colonies became water-logged completely. The hapless and helpless citizens at those places were drawing drinking water from shallow handpumps, and that waters were already heavily polluted due to the seepage from standing water into the shallow aquifiers. Drinking of that polluted water resulted into wide spread cholera and gastroenteritis affecting thousands of people and causing death to hundreds.

However, in India, the River Damodar happens to be the most heavily polluted river which has been turned into a dumping ground for a large number of wastes and effluents. The degree of pollution is so high that the river water does not contain any dissolvent at many places and, as a result, it fails to support the growth of any desirable aquatic organisms. Only organisms such as bloodworms, sludge worms, blood suckers, pathogenic bacteria and some other undesirable animals are found.

The River Mini Mahi in Baroda (M.P.) is also burdened with a variety of industrial and petrochemical wastes. In recent years surveys of several rivers in Madhya Pradesh like the Chambal, the Sona, the Kahipra, the Narmada, the Tapil and the Indravati have revealed that they have been severely polluted with various wastes being discharged from many textile and paper mills, caustic soda plants and distilleries.

The fate of other rivers in many western countries is no less brighter. The River Rhine which literally means "Pure" and flows through the industrial heartland of Europe is dumped with a huge quantity of chemicals at many places along its course at the rate of 25 train-wagon loads every hour. In short, it has become a big sewage drain of Europe, and signboards with "UNFIT FOR BATHING" inscribed on them have been put up at various places all along its bank. Water of the once romantic Seine River in France is now frothy, obnoxious, foul smelling and devoid of marine lives at many places. The St. Lawrence River and the five great lakes of Huron, Erie, Ontario, Michigan and Superior between Canada and the U.S.A. are all heavily polluted. Back home in Manipur, the sight of the Naga and the Nambul Rivers is really pathetic, especially in their segments of Imphal town.

Amongst the various trace metals such as mercury, lead, cadmium, zinc, copper, cobalt, nickel, silver and chromium present in polluted water as a result of the discharge of numerous industrial effluents, the first three happen to be the most mischievous ones from the point of human health. In water mercury slowly combines with organic compounds containing a single carbon atom in each of their molecules in presence of a microbe called methanobacterium formicicum forming a highly toxic chemical known as methyl mercury or mercury monomethylate. According to the wHO, the permissible limit of methyl mercury in water is 0.05 part per million (ppm). However, sea-water around Japanese towns of Minamata and Migata were found to contain this compound as high as 100 times above the tolerable limit. This was due to the poly-vinyl chloride (PVC) plastic factory which was catastrophic. In 1953, many citizens of Minamata were afflicted with symptoms of severe mercury poisoning after consuming fish, crabs, mollusces and shell fish caught from the adjoining sea-water. They complained of severely aching muscles and distorted vision in which they felt as if they were seeing through a long tube. During the following nine years a few hundred people were identified with similar complaints and 36 of them died. Some survivors became blind and deaf soon afterwards and many children were born with improperly developed bones - a congenital disease." In 1965, the same disease broke out again in Miigata.

Besides the manufacture of the PVC plastic, 230 grams of mercury are estimated to be released into aquatic environment from the production of each ton of caustic soda. Discharges from wood pulp and paper industries and pharmaceutical industries producing anti-bacterials like mercurochrome also contain organic mercury compounds. In India survey reports of the sea-water around Bombay, Tarapur and Thane and fish caught from these areas indicate presence of methyl mercury in a deleteriously high level.

Lead, another toxic metal finds its way into the biosphere from various sources, notably automobile exhaust fumes, plumbing, lead smelters and paint industries. 75% of it emanates from gasoline in which an organo lead compound known as tetra ethyl lead (TEL) is used as an anti-knock component. On an average every 3.6 litres of gasoline is mixed with 2-2.5 grams of TEL. As a result of it 300,000 tons of fine lead particles each with an average size of 0.5 micron are released into the air annually. They contaminate various water bodies after coalescing with dust and subsequently being washed down with rain water.

Lead may cause anaemia, fatigue, insomnia, plain irritability, delirium, convulsions, blindness, mental retardation, brain damage, abortion and still births in varying degrees. Lead is known to accumulate inside the body, especially in the bones. There is proven evidence that lead inhibits the formation of a chemical known as delta-amino levulinic acid dehyrase (ALA-dehydrase), an indispensible blood enzyme which helps in the synthesis of haemoglobin the oxygen carrying molecule of the red blood cells or erythrocytes.

Cadmium, a highly toxic element gets gradually accumulated inside the body; especially in the kidneys once it enters the body through inhalation and ingestion. Water is usually fouled by wastes and effluents of electroplasting industries, zinc smelters and certain paint and battery industries. Cadmium is also found to be rich in certain marine lives like crabs and oysters. The maximum tolerable limit of this metal in the human body has been calculated to be 0.01 ppm. It may damage nephrous of the kidneys; cause liver damage and cancer of prostrate; and result in the disease, "itai-itai". Its accumulation inside the body also causes hypertension and heart diseases. Small dosages of cadmium have produced in animals birth defects, brain damage and many other physiological abnormalities.

Frequent testing of nuclear devices by the nuclear weapon countries also causes a grave hazard to human health. Debris of nuclear explosion such as iodine-131, caesium-137, strontium-90, cobalt-60 and left over plutonium-239 leach and ultimately find their way into sub-soil water. They bring about cancers of the thyroid, spleen, liver and blood, genetic defects of the offsprings and many other ailments. Since April 1948, the U.S.A. has carried out nearly 50 nuclear tests on Eniwetok Atoll in the Marshall Islands and that has led to a complete destruction of plant and animal life both on the atoll itself and in adjoining Pacific waters.

Thus the pollution of our aquatic environment is alarming indeed. That is why, sensing the various lurking dangers to human health brought about by it, many of which have already manifested, the U.N.O. created in 1973 a special agency called the United Nations Environment Programme (UNEP) to deal exclusively with this gigantic global problem. In co-operation with other agencies such as the W.H.O., the UNESCO, the International Union for the Conservation of Nature and National Resource (IUCN) and other voluntary agencies, the UNEP is in the forefront to formulate drastic schemes to tackle the problems of environmental pollution and to arouse a dire sense of awareness about it among the masses throughout the world so that science may really serve the mankind and prepare for a really lovable and hospitable biosphere for the future generations to come.

Environmental Concerns

During the last couple of decades, several global environmental concerns, such as depletion of the global protective ozone in the atmosphere, greenhouse effect leading to global warming, acid rains, accumulation of hazardous wastes, desertification and the like, have received increasing attention in all walks of life. Chief among these concerns are the rapid destruction of many others and the consequent loss of genetic diversity. At the same time, it has also been realized that human activities, especially those related to war, industry, modern agriculture and power generation, have been responsible in creating those extremely hazardous situation that threaten not only human life but also the existence of all living creatures on Mother Earth, both through slow, regular poisoning and sudden catastrophic calamities.

One can decipher two opposite views in most of these concerns, related fundamentally to our traditional relationship to the world around us. At one extreme was the notion that nature exists to be used by people, to serve humanity and support civilization. At the other extreme are the 'preservationists' who feel that nature is threatened by humanity, and thus see a need to fence it off in protected areas as far from the influence of man as possible. Such extreme views have characterized both nations and individuals/individual communities in debates as well as actions. However, a new environmental thinking seeks to stake out a middle ground. This thinking sees MAN as the custodian of LIFE on earth, by pointing towards both past and future. This fresh kind of ecological thinking also attempts to bridge the gap in our minds between nature and humanity, between the protected wilderness area and the exploited land-scape. It is in this thinking, one which accepts people's participation in nature as essential at both the local and global spheres. It also actively seeks out ways in which people's participation in nature can be deepened and extended, so that the lovely complex network of biosphere, threatened as it is, can be saved by our best talents.

Keeping the above in mind, it might be possible for us to have glances at global environmental concerns and the north-east's people's participation. One is in fact tempted to argue that because of the new environmental thinking and ethic, there is a positive hesitation of define environmental issues exclusively in the western genre. Further, there is a growing recognition of the right of the poor and tribal communities to firewood, fodder, clean and adequate water and supportive eco-system. Those very agencies that had, at the inception of the environmental debate, posed as leaders in conservation, have now come to be identified as the leading villains, as they were seen to support, often perpetuate, the inequitable social order that was increasingly being reçognised as the root cause of the deepening environmental crisis and the consequent threat of survival.

In the early seventies, there emerged a fresh concern for the future of the earth. This time the critical factor was depletion of "natural resources" and growth of the human population. In 1972, the "Club of Room" published a report where they concluded :

"If the present growth trends in world population, industrialization, pollution, food production and resource depletion continue unchanged, the limits to growth in this planet will be reached sometime within the next one hundred years".

THE WORLD REPORT on environment appeared as "Bruntland Report". This report emanated from the World Commission on Environment and Development, set up by the UN in 1984 under the chairpersonship of Gro Harlem Bruntland, entitled OUR COMMON FUTURE, was published in 1987.

The Commission represented almost all the significant sections of the world, the rich and the poor, the east and the west, the capitalists and the socialists. This report produced that must so far be the most wide-ranging statement on the world's environmental crisis. It mentioned almost all points of view, and showed a rare sensitivity to a surprisingly diverse set of values. However, perhaps because of this very breath, no clear action plan seemed to emerge. Global environmental rhetoric, with this report, had come of age, but global environmental action still lagged far behind.

Hence, neither in India, nor in the rest of the world, has the new order totally replaced the old system. Though it is true that more and more individuals, institutions and governments are being forced to adopt an 'environmental stand', the different interests and perception that have divided the environmental debate. At the fag end of this decade, one can see a world community which is talking more and more about the environmental crisis, but agreeing less and less on what it means, on how it came about and indeed, on the solutions.

Considering human beings are as such creatures of nature as any other animal or any plant, it seems prima facie impulsible that they could deplete nature to a point where all life gets threatened. Surely, the natural order and the laws of nature must have accounted for human beings and their needs.

Plausibly, human beings, as hunter - gatherers of the past were governed by nature's rules. Even today, isolated tribal groups having no contact with the Jaravas of the Andaman Islands, seem to face no crisis of sustainability.

However, an inherent part of modernity, as we know it, seems to involve changes in our interaction with, and use of nature's elements. It seems to involve the colonisation of more and more of the earth, taking it over from other creatures and manipulating it to progressively suit our own requirements. Modern civilization has increasingly discovered methods of immunising itself against the vagaries of nature, of consuming at a growing rate the 'resources' of nature with eyes of unity : treating it as nothing more than one more "raw material" for industries and economic models.

Modern science gives us the arrogance, technology the *means*, increasing population the *alibi*, and poverty the *justification* to destroy nature. Add to these the world's social and economic order enables a few to dispose the many, and you have the recipe for a crisis of sustainability. The present day lifestyle is not sustainable. We are consuming faster than the earth can replenish. The progressive growth of technology and the resultant consumerism and waste have encroached deep into our society. Unfortunately the technology of conservation, regeneration, recycling and frugality are at an infant stage and scarcely available. The force of private profits is still far greater than that of social welfare. Public awareness has not yet been able to make pollution and environmental degradation unprofitable nor conservation and regeneration profitable.

In short, technology that has so far been used to deplete and degrade the earth could now be used to conserve and regenerate it. All that is required is the non-availability of softer, short-term options, and the resultant acceptance of long-term, sustainable, ones.

This, it is argued, would meet the demand for sustainability. But would it promote equity ? Would it not allow a small number of elites to keep their needs and consumption growing, even though regenerative technology kept pace, while freezing a large part of humanity to base, subsistence levels ?

Can the world, especially the poor, afford environmental conservation ? Can they afford the investment that pollution control requires, or the loss of revenue that conservation of forest and other ecosystem implies ? Economic seems to be able to deal with only those 'goods and services' which are, in one way or another, inputs to, or products of, economic activities. Therefore the cost of soil-erosion can only be measured in terms of the resultant loss of agricultural produce etc. Or what is the cost of a tree which cleans air, regulates water-flow, fixes nitrogen that absorbs pollutants, produces biomass and provides habitat to other flora and fauna ?

Such environmental problems also have economic implications which particularly affect the lives of poorer sections of people. Today, 24% of the world population dispose off 81% of the net product in the world. They consume 85% of all energy produced. They also benefit from more than 90% of the overall expenditure on health, education, science and technology. In sharp contrast, 76% of the world population dispose of a mere 16% of the world's net product. More than 2 billion people in the world do not get even clean drinking water, 1.5 billion do not have access to health care and 1.1 billion live below the 'poverty line'.

Poor people living in areas beset with diseased backwardness, hunger and ignorance constitute the supply of cheap labour and their surroundings' sources of cheap raw materials for rich people in developed nations of the world. In many parts of the world, the human population is growing at rates that cannot be sustained by available resources, at rates that outstrip any planned expectations of providing housing, health care, education and energy. In order not to starve to death the poor (who constitute the supply of cheap labour), producers in the developed countries go on violataing nature, which, in the long run, will ruin our entire environment. Thus, in the ultimate analysis, they aggravate rather than solve the food problem. In fact, most development planning displaces poorer, tribal and indigenous people all over the world. For example, the construction of large dams for power generation in India is invariably at the cost of poor tribal people.

To take another example, the dangerous phenomenon of soil erosion grows with desertification as the ultimate consequence. At present 60,000 to 70,000 hectares of forest area is converted to desert every year. No less dangerous for the environment are the measures for reclaiming new areas for cultivation. According to the United Nations Fund for Popular Activities (UNFPA), while the acreage in the developing countries rose by 59 million hectares during the period from 1971 and 1986, the forest areas decreased by 125 million hectares. No doubt, arable soils won this way are biologically poor. They turn into hard dry clay as soon as they loss the protection of the rich foliage in tropical and subtropical forests. At the end, one does not win new agricultural area but loses valuable forests. The EARTH DAY CONFERENCE, 1990, informed that between 1970 and 1980 our world lost 200 million hectares of forest while the desert area expanded by 120 million hectares. Every year, 3.2 billion tons of carbon monoxide are set free or get into the atmosphere because of destruction of forests.

As a result of these processes, the ecological system of the world is endangered. We are threatened by climatic changes with disastrous consequences. Let's listen to the voice of experts : "With the inexorable march of deforestation, the ecological integrity of many areas are disintegrating, causing severe soil loss, aggravating droughts and floods, disrupting water supplies and reducing land productivity" (Postel and Heise 1988 : P). This situation leads to a speeding up of the natural calamities from decade to decade with the resulting increase of human victims (UNDP 1990 : 133). It has also been estimated that the world forest area will be reduced to 12% by the year 2000 A.D. Not the least, each falling tree makes the ecological system of the earth collapse. The critical point will come much sooner. Or has it already come?

Poor people and their communities have been marginalised to an extent that they cannot afford environmental conservation. Their traditional practices of conservation have been evolved and modern methods are not available to them. They also cannot make the investment required for massive projects like pollution control and protection of ecosystems. They rather have to pay for soil erosion in terms of reduced agricultural produce.

Even though a lot of rhetoric has been poured by experts about developing a sustainable environment, it is yet to be translated into action. We do not seem to have realised the cost of destroying our environment. We develop forests and its environment which have taken thousands of years to build their eco-systems ? Can we stop the disturbance to hydrology of an area which regulates climatic condition ? These are the most serious environmental concerns for all of us today.

Now let us have a glance at two contrasting examples from the north-east. Here, ecology exists as a set of tastes and practices, not ideas and percepts. In this ecological tradition, human beings cooperate with nature, by actively participating in it, to produce a richness of ecological variety that would not otherwise exists, while at the same time producing enough for survival. Precisely for this, we have had no beggers in the north-east. In such a tradition, the landscape and the wild undergo continuous but mild change. Thus, it is adoptable to all sorts of minor ecological alternation, nevertheless, it works to conserve, to protect and conserve what is needed and what is good, it maintains itself.

A unique example in this regard is this Dzuku valley tucked away in the remote northern most corner of Manipur. We are drawn to Dzuku because they provide a concrete, non-professional example of our new environmental thinking. Dzuku valley is a rare, unique ecosystem that harbours not only a spectacular variety but it is also an example of the local people's traditional experience and expertise to develop creative solutions to environmental problems. For the people here, conservation is all action, a way of life, but perhaps no debates, no programmes. They practise conservation, in nature, with nature, blissfully free from academic jargon.

Environmental Concerns in Manipur

Manipur in geographically located in a zone where the world's heaviest rainfally is recorded. At present, the State receives 142 cm. rainfall annually. It receives rains during the summer monsoon, from the middle of June through to September. During these four months, it receives 60% of the total annual rainfall. However, because of various environmental problems, the state faces the problem of water i.e. droughs in lean seasons and flood in the peak season. There are only two river systems in the State. Both of them are exclusively rainfed with the torrential headstreams of Barak and Chindwin (Burma) river systems. In olden days tributaries of both the drainage systems were perennial. However, in the present, even the main headward streams of the Barak and the Manipur River (a tributary of Chindwin) are in lingering conditions. Most other tributaries and the numerous contains and springs have become seasonal.

This occurrence of seasonality in the drainage system of the State has significant correlationship with the increasing trend of deforestation in the catchment area. The rainfed headstreams of the Barak and the Manipur River, we are afraid, would become seasonal if the current rate of destruction of forests continues. In fact, these forests covering the catchment areas contribute to the storage of water which feed the rivers the year-round. The conditions obtained in the catchment areas of about 12,773 sq.kms. are precarious. The luxuriant forest growths of the catchment areas have been mindlessly exploited by both Jhumming and logging. Annually, more than 950 sq. kms. of forest area is being transformed into Jhum fields. Over and above, the increasing growth in population has not only added a tremendous pressure on land under cultivation, but it also has shortened the Jhum cycle from 10-20 years to 3-4 years in the hills and uneconomical land holding in the valley. The expansion of jhumming areas means extension of deforested areas. Jhumming clearance of forests accelerates surface run-off, soil wash/erosion, siltation in river bed and lakes in the valley causing flash floods and droughts alternately.

It is worth mentioning here that there were about 500 lakes and wetlands in Manipur valley. Now, most of them have become shallow or have dried up and are used for human habitation. Special mention needs to make here of the Loktak lake, the biggest fresh water lake in NE India. Since the last couple of decades it has been consistently shrinking. Within the last one decade, 30% of the water area of Loktak lake has already shrunk. Despite huge investment of money in the Loktak Multipurpose Project, the average projected land to be brought under irrigation is marginal and power generation has not proved to be profitable in the context of year-long load-shedding (ranging from 3 to 10 per day) in the State. High siltation rate in the lake further poses a threat of shortening of its longevity.

As a matter of serious concern, it must be noted that Loktak provides us with a unique example of a rich wetland eco-system, unfortunately of the past. In recent years its eco-system and that of the entire valley of Manipur have deteriorated drastically. Without any shred of doubt, it may be mentioned that the construction of Loktak-Hydro Electric Project and the controversial Ithai Barrage have been almost entirely responsible for its misearable deterioration, to an extent that even a National Park cannot provide any safe-guard to Sangai, the only surviving species of brow-antlered deer in the world.

The negative impact of the degradation of Loktak lake is clearly discernible from the predictment of the fishing community. For the normal functioning of Loktak Hydro Electric Project, the water level of the lake has to be maintained at a particular level throughout the year. Resultantly, it has caused serious difficulty and misery for the fishing community to concentrate in the peripheral villages of the lake. At one stroke, they have been deprived of their traditional techniques of fishing. There is also enough evidence of the misuse of heavy doses of insecticides and pesticides like Thionel-Endosulfan, Endosulfan-Technical, Pestoban 20 and Rogo 300 which are being committed in the fresh water for the purpose of catching fishes. The fish thus caught are marketed in Manipur, causing health hazards and occasional deaths.

Additionally, in the villages on hills and foothills, women and children spend a large part of their time travelling long distance to collect firewood and water. In some of the hill areas, the women folk walk at least ten kilometres, three out of four days, for an average of 7 hours per day, to fetch 25 kg of wood in each headload. Children of the area become the unwitting victims of the continuous energy and water hunger in the villages on the hills. Most of their time is taken up by tasks of water and fuel gathering. In small towns poor families spend a much larger proportion of their income on getting their energy needs, often as much as 15 to 25 per cent of their total income. Given the fire wood that these poor families need to spend 80 to 90 per cent of their income on food, rising fuel costs automatically leads to reduced food intake. For an already malnourished population, this indeed is not simply unfortunate but a threat to their right to life. Cooking energy shortage not only creates serious nutritional problems, it also leads to increased incidences of health hazards and epidemics. The widespread incidences of scabies in India can certainly be attributed to inadequate fuel to heat bath water during the chilly winter which forces many poor people to avoid baths for a long period, leading to scabies and other skin diseases. In Manipur, it has been confirmed that 75% of the health problems and diseases (diarrhoea, digestive and intestinal, malaria, skin etc.) are water borne. Most of the problems outlined above are virtually directly related to the degradation of environment in our small State.

Futher, mention may be made of the 161 mere TIPAIMUKH HIGH DAM which is being constructed at the confluence of Tuivai and Barak rivers bordering Manipur, Mizoram and Assam. Even a cursory glance at the Tipaimukh High Dam (THD) Project Report indicates how the entire developmental planning and implementation in our country is irresponsible and counter productive. This dam will displace thousands of tribal families of about 60 villages in Churachandpur and Tamenglong districts of Manipur. The THD will submerge an area of 286.36 sq.kms in Manipur alone. What is of utmost concern to us is that the construction of this dam has been going on for more than half-a-decade but no environmental impact assessment has been done as yet. In fact, since the 60s, at least 4 dams on these rivers were proposed and planned but after the initial geological and environmental studies they were abandoned. In the case of THD no such studies have been carried out. The Project Report of THD very casually states that there is no worthwhile flora and fauna in the submergeable area.

Since the THD is being constructed without any survey whatsoever and without any environmental impact assesment, the common tribal people of the areas as well as concerned scholars, social and environmental activists have protested against the construction of the dam. The Zeliangrong Students Union and the Hmar Students Association have also represented to the Prime Minister of India to review the construction of the THD. A group of elderly statesmen, scholars and activists formed a group called Action Committee for Tipaimukh (ACT) to seriously look into the matter. Some members of ACT went on a fact-finding mission on THD towards the end of 1991. To their utter shock and surprise,

ACT found that the location of the dam itself is on the Taithu Fault line. In fact, the Barak river runs under a number of fault lines - there are about 100 faults in this river system. The surrounding area of the proposed dam, on the Manipur side, receives very high rainfall. The small valleys in the area will be severely affected because of the rise in the height of the beds of Tuivai, Barak and Tuivang rivers. Most of the valleys in the area are narrow and gorgelike.

According to a Geological Survey of India report, the 75 km. radius of Tipaimukh is prone to earthquakes that may measure upto 8 or more on the Richter scale. The region has already experienced two major quakes of 8+ magnitude within the short span of the last 50 years. The next predictable quake may occur at around 2010 A.D.

ACT also discovered that the statement that there were no worthwhile flora and fauna in the submergeable area as mentioned in the THD Project Report was totally false and baseless. The members of ACT collected firsthand, primary information in this regard during their fact-finding visit in 1991. The flora of the area includes various types of orchids and endemic medicinal plants. The hill ranges in the area are very steep and largely covered by forest and orange-lemon orchards. With regard to the fauna of the area, special mention may be made of rare species like reptiles, pythons, crocodiles, turtles and amphibians, gibbons and various deer species. Various species of fishes and birds (including migratory ones) were found during the ACT survey. On the whole, the construction of THD poses a major threat to the environment and ecology of the entire state of Manipur.

Jhumming Cultivation in Manipur and its impact on Environment

Manipur with a total geographical area of 22,327 km^2 has a total population of 1,827,000 who live in 2035 villages and 32 towns and cities as per 1991 census. Thus Manipur has a density of 82 persons per km^2 as against 264 in Assam, 262 in Tripura, 78 in Meghalaya, 73 in Nagaland, 33 in Mizoram and 10 in Arunachal Pradesh. All these are lower than the Indian average of 267 per km^2 which is exceeded only by Assam in the N.E. Region of India.

In Manipur as per 1981 census 1.25% are scheduled castes and 28.55% are scheduled tribes and thus these two put together account for 29.80% against the Indian average of 15.75% S.C. and 23.51% S.T. population. Thus the SC & ST population of Manipur is less in percentage than that of Indian average figures. In Manipur 1.9 lakh population lived below the poverty line in 1983-84*[1] which is 22.97% of its total population against the Indian average of 32.11%. Thus Manipur stands at a better level than that of Indian average so far as general poverty is concerned. This is mainly because of the lower percentage of SC and ST population, compared to Indian average. Thus to sum up Manipur's density of population is lower than the national average, have less percentage of people under SC & ST population, less percentage of people who live below the poverty line than that of India as a whole. But, the danger lies in its population growth rate which is 28.56% during 1981-91 as against 23.50% during the same period in India. Thus the population growth rate per annum works out to 2.85% in Manipur as against the average annual Indian growth rate of 2.35%. This is an ominous sign, which if not checked, will eat up all the plus points of Manipur in the future decades of India and she will be lagging behind the economic growth of India and more and more people will be pushed to below the poverty line.

To pull up Manipur in the ladder of economic growth, Manipur should arrest its fast degrading environment and eco-system in the hills, valleys and wetlands so that its economy does not sag or even collapse by the turn of the century or by 2011 A.D. by the sear weight of her fast growing population which is in a slightly better position than that of the Indian average.

Shifting Cultivation (Pam Lou)

Here an attempt will be made to evaluate the extent & intensity of shifting cultivation as the root factor for eco-system degradation in Manipur. In case of shifting cultivation, popularly known as Jhumming (Pam lou in Manipuri) a part of the forest is selected by the community during winter in the months of December & January. The whole community moves to the site with "daos" (Thangchau in Manipuri) and axes. Trees are felled and left to dry till the month of March when they are set on fire. The ashes are mixed with the soil which is loosened with the help of ordinary spade, dao, hoe, drilling road, etc. Then the field is ready for sowing. This is considered as the first step in the transition from food gathering and hunting economy to food production.

All activities connected with this type of cultivation are performed mostly by community labour of the village in Manipur hills. The hill slopes which are to be cultivated are chosen at a village meeting. According to the needs and consent of the community the strips of land are paralleled out among different families in a village held. But as long as a person is in cultivating possession of certain strips of land he is recognised as the owner of those plots. As soon as these plots are left to rest, they become the property of the community.

With the onset of the monsoon by mid-June, torrential precipitation washes down the ashes into the cultivated fields at the foot hills down below. Thus, the jungle clearing involves cutting down grass, shrubs, trees and the vines. Shifting cultivation is a system to the cultivation of paddy. Crops are given under rained conditions during Kharif season and they are harvested in winter. These places of land are used for two or three years when successively different crops are, grown. Then they are abandoned so that they could recuperate. This practice of growing mixed crop by the tribals is dictated by their food habits and ecological conditions.

The recuperative cycle varies from region to region. The shortening of the cycle of rotation has caused extensive damage to the natural vegetation in the hills of Manipur. This practice of cultivation has completely destroyed the forest coverage and the soil has been seriously impoverished and in consequence many hill areas of Manipur have been out-cropped and they have become unsuitable for bearing any vegetation. With the approach of this stage the warning signal is struck for the Jhum cultivators to shift for a new place of virgin forests to start the jhumming afresh. Such type of cultivation is considered devastating and disadvantageous. Some of the evils of this type of cultivation are as follows :-

(i) it causes the loss in edaphic fertility.

(ii) causes erosion of thin top soil and deterioration of soil.

(iii) ruins valuable timber for the sake of much less values crop or grain.

(iv) causes very heavy floods in the down streams of water channels endangering life and property and the end product is degeneration of environmental and ecological balance which is almost a permanent damage.

Thus shifting cultivation affects the entire economy of Manipur State as a whole.

The shifting cultivation has also another problem for consideration. Due to repeated cutting and burning of forests, most of the tracts used for shifting cultivation have become barren. With the increase of tribal population the need for larger patches of land for jhum cultivation is being needed every year. Thus year after year, jhumming areas are expanding on the Manipur hill ranges with the increase of tribal population. It is true for the hill slopes and mountain ranges of the Arakan Yomas, demarcating the border between India and Myanmar (Burma) on the eastern side of Manipur State and on the western border of the valley, the Thangjing range and the Koubru range suffer from the same problem. These hill slopes and ranges have now become unsuitable for the growth of the forests. Large-scale conversion of such tracts into reserved areas does not help very much in augmenting the forest wealth. In fact, what happens as a result of jhumming is that the tribals are deprived of their only means of living and are forced to lead a life of poverty and misery.

Based on these reasons, the State Government have to ban the shifting cultivation and implement the policy of preservation and conservation of the natural forests of Manipur.

What is, therefore, necessary is that before banning the shifting cultivation, sufficient care should be taken to provide the tribals with good land and rehabilitate them in colonies, established in suitable sites either in the valleys or in the foot-hills near about their parent habitats.

In the very first a few showers finger rills appear in multiple slopes and with the advancement of the S.W. monsoon these finger rills become rills and also multiply in number and with increased depth. At the end of the monsoon, minor rills almost become gullies through which about 1/4 of the thin top soil is washed down the slopes in numerous water channels. At the maximum height of the jhumming the bed rocks are exposed in the very first season of cultivation & by the end of the third year of jhumming, the entire top soil of the hill slopes is washed down through streams and the bed rock is fully exposed in the entire area, which once was the home of lust Tropical Wet - Ever Green forest. With the process of deterioration of soil, the already small per hectare yield, becomes minimal and the tribals are forced to shift in quest of new areas to repeat the same process once again in virgin forest region. The abandoned Jhum area is left fallow where the original forest never regenerates again and they become barren where either scrubs or minor trees grow in a scattered manner. Thus the entire eco-system of mountain slopes changes in both flora & fauna. Thus the Pam Lou cycle has continued in the past and it is continuing even today unabated.

Down the water channels the ripples of shifting cultivation is being felt in a still worse manner. The minor and medium irrigation projects get silted up & their life span is reduced drastically & all the previous efforts to irrigate and generate hydel power are reduced and ultimately nullified. Further down below, in the wet lands and lakes in the Imphal valley bear the worse brunt of Jhum cultivation. They are silted up, their capacity to hold water is reduced, lake flora & fauna are adversely affected due to addition of silt and lowering of the depth of water.

As a result, about 11 shallow lakes and swamps in the Manipur valley have dried up. Most important of them are Lamphel Pat, Kakwa Pat and Porom Pat and are now being used for human settlement and cultivation. Currently about 23 shallow lakes are on the verge of drying up. Amongst them Waithou, Loushipat and Poiroupat are most important. Even Loktak Lake, the largest fresh water lake in the N.E. India, has shrunk by about 30% in its water holding capacity during the past decade. As a result, now if there is a continuous rain for three days in the hills on the fourth day, the Manipur valley is flooded, a phenomenon earlier unknown in the Seventies. The flood is also affecting Imphal City itself. This is happening due to the shrinkage in the water holding capacity of the Loktak Lake due to massive siltation. The net result is, the frequency of occurrence of such floods has increased to almost an annual phenomena. The Manipur Valley has experienced devastating floods in 1985, 1986, 1987, 1989 and 1990.

Thus, while in the valley and wetlands the problem is flooding in the hill tops and hill slopes, the problem is scarcity of water, although the State as a whole receives an annual average precipitation of 142 cms. 60% of which is received during the S.W. monsoon four months. In the water shed areas of the Barak, Manipur and Chindwin rivers, the precipitation is as high as 130 cms. to 220 cms. per annum.

About three to four decades before most of the tribal villages were located at the highest points of the hills and mountains but due to degradation of the soil, these high landers have been compelled to migrate downwards, where the perennial water points are still available and the tribal ladies are daily travelling 5 to 10 kms to fetch a pail of water for their domestic consumption.

It is established that annually more than 600 km^2 of forest area are being transformed into jhum fields. Due to continuous growth of population and tremendous pressure on land, the old Jhumming cycle of 10/20 years has been reduced to only 3/4 years. The extension of areas under Jhumming means shrinkage of forest land.

Jhumming has also its impact on the river channels. Two decades before both Barak and Chindwin river systems were perennial but in nineties, leaving the main First Order streams of these two rivers all others are in a lingering condition and most of the higher order tributaries (as per British system of stream ordering) have become non-perennial. This process has now engulfed all the shore regions of wetlands, leaving only the middle areas intact.

But the challengers of the day is to bring a halt to this pernicious process of shifting cultivation and if not halting the process, at least it must be slowed down, so that these wetlands and water channels continue to function normally for a longer period and thereby lengthening the period of getting economic benefit from them.

Exo-system problems posed by shifting Cultivation

Eco-system of virgin soils is completely disturbed in the

process of Jhumming. It never comes back to the old eco-system during the fallow period since drastic structural changes take place in the odaphic conditions during the 3 to 4 years of jhumming.

In these processes many species of flora and fauna belonging to the minor stratum, have disappeared. Thus a truncated eco-system has emerged which is highly injurious in the long-run.

The data have also not been generated as regards the quantum of soil erosion taking place on the high slopes as a consequent of jhumming. This is a major gap which is sought to be measured in case of Manipur at various micro-watershed management regions.

The concept of watershed management involved optimum use of natural resource of land and water within a natural physiographic area and recommended by the Central Team constituted by the Government of India, for study of the catchment area that terrace cultivation in hill slopes are suitable alternative to shifting cultivation in some parts of Manipur.

Effect of Jhumming on Water Bodies

Shifting cultivation is preceded by cutting and burning of the vegetative cover. Jhumming not only destroys the existing delicately balanced eco-system in the virgin forest of mountain slopes but also it initiates and accelerates soil erosion which in turn adversely affect the environment and eco-system of the water bodies of Manipur. The river Thalweg (valley long profile) is not only checked up as a consequence of extra load to be carried down streams while volume and velocity of water remain constant. Thus new gradients are produced in the valley long-profile, particularly when rejuvenation of the valley long-profile takes place at lower reaches as a consequent of the deposition in the water channels of the valley trunk streams. This process is the starting point of first deformation at the meeting zones of the wetlands (lakes) and surrounding hill slopes.

These process disturb completely the environment and eco-system of the water channels which was originally unaffected, by the excess amount of loads, before jhumming in the catchment areas while volume and velocity of water remain constant.

The inherent potential of land and water resources in the area, is governed by a set of factors such as climate, geological base, topography, edaphic characteristics, present land use, and socio-economic levels. An optimum interaction amongst these factors can be observed within a watershed and can be employed specially for proper utilization of land and water resources to prevent eco-system and environmental degradation of land on the hill tops and hill slopes.

Finally, the beds of the wetlands (lake etc.) are being continuosly silted up at an accelerated rate and thereby decreasing the water holding capacity of the wetlands and its longevity. This process affects the lake eco-system in the following manner :-

1. As a consequence of silt deposition the bottom topography of the lakes is changed.
2. Many micro-organisms living in mud deposit at the bottom bed of the lakes have been adversely affected. The flora and fauna of the lakes are also changing as a consequence of rising of the bottom bed levels.

The above factors have affected the fish population adversely both in quality and quantity. This happens because of changing wet-bottom and composition in the lake beds.

Thus eco-system of the stagnant wetlands like Loktak lake, Ikop, Kharungpat, Pumlen etc. are very fastly degrading.

Discharge Data of the Rivers

The striking feature in the drainage system of Manipur is the dominant control of the Manipur river and the Barak river. The rivers of the state belong to three drainage systems. These are :-

(1) The Manipur river system

(2) The Barak river system &

(3) The Chindwin river system.

Both the Chindwin & Barak rivers have originated from the hilly terrain of the State where jhumming cultivation is in vogue. The Chindwin has a number of head-streams like Iril, Imphal, Sekmaijin, Chakpi, Khuga, etc. popularly under the system of Manipur River. The Barak system consists of five main rivers viz., Barak, Makru, Irang, Tuivai and Jiri rivers.

The Manipur River System

The eastern half of the State has been predominantly drained by the Manipur River (headstreams of the Chindwin) and its tributaries. The drainage system of the Manipur River consists of two sub-groups of tributaries i.e. (a) tributaries directly discharging to the Manipur river, and (b) those into the Loktak lake and other lakes of the valley.

(a) Imphal, Iril, Thoubal, Chakpi, Khuga belong to the first group. (b) While Nambul, Nambol, Thongjaorok, Ningthoukhong, Yangoi Achouba and Yangoi Macha and other 4 or 5 rivuletes and streams with a combined discharge basin of about 650 sq. km. flow into the lakes mentioned above. It is interesting to note that a channel called Khordak, emerges from the Loktak Lake and joins the Manipur River, which is so far fed by the first group of the tributaries in the upstreams. From the point of confluence of the Khordak and the Manipur River, the single river system commanding almost the whole of Manipur emerges.

The valley long profile of the Manipur River shows that it runs through a smooth gradient from its origin in the north (Ukhrul and part of the Senapati Districts) sown to the Imphal town. In this part the Manipur River is known as Iril. The Imphal and the Iril — these two big rivers of the valley meet at Lilong town. From the confluence the course of the Imphal river is 75 km and that of the Iril is 120 km. The Thoubal river joins the main rivers near Mayang Imphal town. It has a total length of 105 km. All these three rivers have a combined catchment area of about 2,600 km^2. The Thoubal river has a steep gradient and brings a huge load of sediments to the Manipur River and is responsible for occasional devastating floods in the valley. Further south, the Manipur River receives two of its big tributaries coming from the opposite direction i.e. from the southern direction. The Khuga river meets it near Ithai. It has a course of 80 km. It is a mountain stream of steep gradient, and during the rainy season, brings so much of course traction load and water that the normal flow of the Manipur River is disturbed and even blocked temporarily. Although the level of the Manipur River bed is lower than that of the Loktak water level, yet during the flood season, such a huge amount of water is impounded in the river that there is a reversal of flow of water from it to the Loktak Lake comes down to the level of Manipur River through the same Khordak. Thus Khordak acts as a communication channel in between the Manipur River and the Loktak Lake. On the other hand the Loktak Lake acts as a temporary natural reservoir for extra flood water.

South of Sugnu township the Chakpi river joins the Manipur River. It originates from the southern hills, describes a course of 60 km and commands a drainage basin of about 800 sq. km. Being a mountain stream it has a steep gradient and brings huge load of debris to the Manipur River. After the confluence with Chapki, the Manipur River flows past through a long deep gorge and finally joins the Chindwin River below the town of Kendet in Upper Burma.

The Barak River System

The Barak system consists of mainly seven tributaries apart from Barak itself. They are as follows :-

1. Makru 2. Irang 3. Tuivai
4. Jiri 5. Tipai 6. Leimatak and
7. Maklang

The Barak, Makru, Irang and Jiri originates from the southern slopes of the western Manipur ranges (covering Senapati/ Churachandpur Districts) while Tuivai, has its origin in Myanmar and is joined by the Tuila, Tuilak and Tuisa rivers before its confluence with Barak at Tipaimukh. The system (Upper Barak) covers a drainage basin of about 9,518 sq. km. The hilly terrain of Upper Barak, with steep gradient has good drainage conditions in contrast with the lower reaches (flood plains of Cachar District of Assam) where flat ground with many meanders create drainage congestion.

The Chindwin River System

Tributaries are :

— Akonglok
— Chingai
— Chamu
— River yu
— Maklang

— Tuyangbi
— Lokchao and
— Lalimlok

The appropriate catchment area of above rivers are given below :-

	River	Catchment area in km²	Percentage of the total area
1.	Barak	8,700	38.8
2.	Manipur	5,600	25.0
3.	Chindwin	8,100	36.2
	Total	22,400	100.0

Out of the total estimated precipitation of 18,487 million hectare metres, about 80% of it flows downstream channels as surface runoff due to declining forest coverage and loss on evaporation, leaving only 20% for the use in hill tops, valleys and in plains.

Thus the best way to increase the ground water recharge is to provide good forest coverage on the hill tops and hill slopes with permanent grass cover, coming as close as to conserve water resources is to conserve the catchment areas.

Conservation of Water Resources in Manipur

Introduction

Manipur, the State of North-Eastern India lies in one of the heaviest rainfall regions of the world. It is situated at an approximate aerial distance of about 370 km. in the north-eastern direction from the Bay of Bengal. Yet, it is essentially within the tropical climate modified by local factors like site, situation, altitude, topography, direction of the prevailing local winds and seasonal rhythm. The area has a tropical location in the midst of the Eastern offshoot of the Himalayas. Physiographically, the State may broadly be classified into two regions viz., (1) Hilly terrain and (2) the valley. The valley contains a large number of lakes and swamps while the hilly terrain regions are netting with the catchment areas of the two river systems of the region (the Barak and the Head Stream of the Irrawadi).

Water Resources

The main source of water in the State is mainly surface water i.e.lakes and rivers etc. The total surface water resources of the State has been roughly estimated as 18,487 million hectare metres in the form of average annual yield. However, the underground water resources of the state is yet to be explored.

Almost all the surface water were supplied by the rain water. Manipur State receives most of her rainfall during the monsoons. The monsoons are seasonal winds blowing into India from the surrounding oceans (Arabian sea and Bay of Bengal) during the summer and blowing out from the Indian subcontinent into the ocean during winter. The monsoons are south-west monsoon (summer) and north-east monsoon (winter).

The entire State receives rain during the summer monsoon between the middle of June and continue upto September. It receives more than 60 per cent of its annual rainfall during these four months. The annual rainfall in the State is 142 cm while for the National record is only 117 cm. The distribution of rainfall over the state depends largely on the orographical features.

As in other parts of the country, the rainfall in the State is strictly seasonal. The people have to collect and store the water for use during the rest period. It is this aspect which creates the problem of even drinking water in all parts of the State. Even in Cherrapunji the heaviest rainfall (1,100 cm.) area of the world suffer during the spring and pre-monsoon months due to lack of storage facilities and lack of awareness in conserving the available water resources.

The current trend of rain in the State is not one of continuous. Bursts of general rain alternate with breaks and dry spells. This characteristic of rainfall is one of the most important features of monsoon. The commencement of the rains may be considerably delayed. The rains may terminate earlier than usual or scanty in this part of the country. Abnormality in rainfall (heavy or scanty) is the only ultimate cause of local flood and drought in the State. It is really unfortunate, the State has singular habit of not making use of its available water resources. Even though by 2001 A.D. Manipur will use only a tenth of the total annual rain, it is expected to face the threat of shortages in the decades from now. The reason for this is simply that we cannot hold on to all the water it receives. More than 80 per cent of the monsoon water disappears through surface runoff and evaporation etc.

Forest and Water

Over 60 per cent of the State's average annual rainfall of 142 cm comes in about four months; the south-west monsoon (June-September) brings in two-thirds of the water. Throughout the state, there is only two rivers and both of them totally rainfed torrential head streams of the Barak and the Chindwin River systems. In the olden days tributaries of both the drainage systems were perennial. Nowadays, even the main headward stream of the Barak and the Manipur river are in lingering condition. But most of the tributaries along with enormous fountains and springs are seasonal.

The occurrence of seasonal character in the drainage system of the state has a significant correlationship with the increasing rate of deforestation in the catchment area. The rainfed head stream of Barak and the Manipur would be seasonal were it not be for the forests covering the catchment areas which store water and feed the rivers the year round. However, present conditions happening in the catchment area of about 12,738 sq. km. is a precarious one. All the luxuriant Forest growths of the catchment area is continuously exploited both by the Jhum cultivators and timber exploitations. Annually more than 600 sq km of forest area were transformed into Jhum-field. Over and above, continuous growth of population add a tremendous pressure on land and consequently Jhumming cycle has now been shortened from 10/20 years to 3/4 years. The expansion of area under Jhumming means extension of deforested areas. Jhumming clearance of forest area accelerates surface run-off, soil wash/erosion, siltation in river beds and shallow lakes in the valley causing flash floods in the rainy season and drought in the lean seasons.

Forest has a significant role in protecting flood occurrence and prevailing of drought condition. Forest provides a cover that cushions the impact of rainfall on the hill slopes, thus facilitating infiltration of water into the soil. As a result cold transparent water is releasing through perennial fountains and springs as a main source of a river water. However, the present situation in the state is a precarious condition. Just after the rainy days all rainwater were free to flow in the form of surface runoff, evaporation and evapo-transpiration. As a result of this another threatening effect on water is the problem of drought in the state. Now a days, the late arrival of monsoon causes drought conditions in the valley areas particularly at the time of cultivation of crops. This phenomenon has become a regular feature in the state.

The problem in the hill area may be reflected as a scarcity of water. Thirty/Forty years back most of the tribal villages were located at the highest point of the hills and mountains. At present those highlanders are compelled to migrate downward where the perennial water points are available. Earlier when the hill ranges were thickly clad with green vegetation the water sources were available even upto the highest peak. Now, the situation is no longer there.

Conservation of Water

It is obvious that the first step in any rational water management or conservation policy must be soil conservation. To retain more water then it does at present, the ground must have vegetation on it, and for this the soil must be healthy. The best way of increasing ground water recharge is to provide a good forest cover, with permanent grass cover coming a close second. Collecting water is equally important. Perhaps the most efficient and economical method of conserving water resources is to protect the catchment area of the major drainage systems of the state.

Problems of Flood & Drought in Manipur

More than any other surface process, running water is responsible for sculpturing the landscape. Rivers erode their valleys and, in the process, provide a system of drainage that is delicately balanced with the climate, topography, rocks and soil. Rivers are also basic transportation system that carry rock and soil material delivered to the channel by the tributaries and the slope processes on the valley sides. Much of the sediment transported in rivers is periodically stored by deposition in the channel and on the adjacent flood plain. These areas, collectively called the reverine environment, are the natural domain of the river. Lateral migration of bends of rivers and over-bank flow is termed as flooding. Channel discharge at point where water overflows the Channel is called the flood stage and may not coincide with the property damaged. The term flood frequently connotes the stream or river that has reached a high water condition likely to cause damage to personal property on the valley/plain. This is obviously based on a human perception of the event and, therefore, will change or vary as an human activity on the reverine environment or the regime of the river. Most river flooding is a function of the total amount and distribution of precipitation and the rate at which it infiltrates into the rock or soil and the topography and also caused by the destruction of forest in the catchment area.

The occurrence of flood and drought in the State needs a behavioural analysis of the following drainage system along with the rate of precipitation including amount of evapo-transpiration, retaining capacity and the current practices of land-use in the catchment area. The drainage system of Manipur may be grouped into two viz., (1) The Chindwin system and (2) The Barak system. Both the rivers originate from the hilly terrain of the State where Jhumming cultivation is in vogue. The Chindwin has a number of headstreams like Iril, Imphal, Thoubal, Sekmaijin, Chakpi, Khuga, etc. popularly under the system of Manipur River. The Barak system consists of five main rivers viz., Barak, Makru, Irang, Tuivai and Jiri rivers.

1. The Chindwin System : The eastern half of the State has been predominantly drained by the Manipur River (Headstreams of the Chindwin) and its tributaries. The drainage system of the Manipur River consists of two subgroups of tributaries i.e. (a) tributaries directly discharging to the Manipur River, and (b) those into the Loktak and other lakes of the valley. Imphal, Iril, Thoubal, Sekmaijin, Chakpi, Khuga belong to the first group, while Nambul, Nambol, Thongjaorok, Ningthoukhong, Yangoi Achouba and Yangoi Macha and other 4 or 5 rivulets and streams with a combined discharge basin of about 650 sq. km. flows into the lakes mentioned above. It is interesting to note that the channel, the Khordak, emerges from the Loktak lake and joins the Manipur River, which is so far fed by the first group of the tributaries. From the point of confluence of the Khordak with the Manipur

River the single river-system commanding almost whole of Manipur emerges.

The long profile of the Manipur River shows that it runs through a smooth gradient from its origin in the north (Ukhrul and part of the Senapati Districts) down to the Imphal town. In this part, the Manipur River is known as Iril. The Impahl and the Iril - these two big rivers of the valley meet at Lilong town. From the confluence of the Imphal and the Iril upward the total length of the Imphal river is 75 km and that of the Iril is 120 km. The Thoubal river joins the main rivers near Mayang Imphal town. It has a total length of 105 km. All these three rivers have a combine catchment area of about 2,600 sq. km. The Thoubal river has a steep gradient and brings a huge load of sediments to the Manipur River and is responsible for occasional devastating floods. Further south, the Manipur River receives two of its big tributaries coming from the opposit direction i.e. from the southern direction. The Khuga river meets it near Ithai. It has a course of 80 km. and a drainage basin of about 620 km^2. It is a mountain stream of steep gradient, and during the rainy season, brings so much of coarse sediments and water that the normal flow of the Manipur River is disturbed and even blocked temporarily. Although the level of the Manipur River bed is lower than that of the Loktak water level, yet during the flood season, such a huge amount of water from it runs into the Loktak lake through the Khordak channel. When the Khuga recedes and the Manipur River reclaims its southward flows the excess water of the Loktak comes down to the Manipur River through the same Khordak. Thus Khordak acts as a communication channel in between the Manipur River and the Loktak lake. On the other hand the Loktak lake acts as a temporary natural reservoir as that of the Tole Sap of Kampuchea (Cambodia) which acts as a temporary reservoir for the flood water of the Mekong River.

South of Sugnu township the Chakpi river joins the Manipur River. It originates from the southern hills, describes a course of 60 km and commands a drainage basin of about 8002. Being a mountain stream it has steep gradient and brings huge load of debris to the Manipur River. After the confluence with Chakpi, the Manipur River flows past through a long deep gorge and finally joins the Chindwin River below the town of Kendet in Upper Burma.

2. The Barak System : The Barak system consists of five main tributaries namely, Barak, Makru, Irang, Tuivai and Jiri rivers. The Barak, Makru, Irang and Jiri originated from the southern slopes of the western Manipur ranges (covering Senapati/Tamenglong/Churachandpur Districts) while Tuivai, has its origin in Burma and is joined by the Tuila, Tuilak and Tuisa rivers before its confluence with Barak at Tipaimukh. The system (upper Barak) covers a drainage basin of about 9,518 km^2. The hilly terrain of Upper Barak, with steep gradients has good drainage conditions in contrast with the lower reaches (flood plains of Cachar District of Assam) where flat ground with many meanders create drainage congestion.

Land-use and Precipitation factor for causing Flood and Drought in the Region

General Land-use studies of the State are being carried out based on topographic maps, field visit, existing records and aerial photography. The present system of land-use particularly in the catchment/watershed areas of the two systems of rivers i.e. the Chindwin and the Barak is totally based on shifting cultivation. This system of land-use is popularly known as JHUMMING (pam lou in Manipuri) and is considered as the first step in the transition from food gathering or hunting culture to food production. It is a cyclic cultivation in the tropical and sub-tropical regions. The most primitive practice is still being in vogue throughout the hilly tracts of Manipur. Continuance of this Jhumming practice has an adverse effect on the balance of the environmental eco-system of the region. However, such kind of practices has a close link with socio-economic and cultural factors including lack of communication leading to physiographical remoteness and isolation. The Jhumming cycle in the same plot of land which extended to 20-30 years in olden days has now been shortened to 3-4 years because of pressure on land due to increase in population and decrease in productivity leading to extension of more areas under Jhumming. The extension of areas under Jhumming means extension of deforested areas. Thus, the area of fresh clearance of forest every year in Manipur is estimated to be 600 km^2. Jhuumming clearance of forest area accelerates surface run off, soil wash/erosion, siltation in river beds and shallow lakes in the valley causing flash floods in the rainy season and drought in other seasons. It is interesting to note that the State belongs to one of the most heaviest rainfall regions of the world. The State, particularly in the watershed areas of the two rivers system stated earlier is receiving annual precipitation of 1,300 to 2,283 mm., spread over seven months from May to November. Over and above, most of the watershed/catchment areas of the river are exposed to weathering agencies. The damage caused by raindrops hitting the exposed, weathered hill slopes at a high velocity is the first in the erosion process. We may think of the raindrops as miniature bombs hitting the soil surface. Experts have estimated that more than 100 tons of soil per acre may be detached by the raindrops in a single rainy season. Consequently the splashed soil is moved downward silting the river bed and shallow lakes and swamps in the valley were dried up and used for human settlement. Currently about 23 shallow lakes were at the verge of drying. Among them Waithou, Loushipat and Poiroupat may be mentioned. Even the Loktak Lake (the biggest fresh water lake in the North Eastern India) has been shrinking by about 30% of the water within a decade. On the other hand if there be a continuous raining for 3 days in the hilly arrea there will be a flash flood in the valley of Manipur on the 4th day. Hence, there was an occurrence of flood consequently in the year 1988 also. These are the only product of excessive destruction to forest in the catchment area.

Forest has a significant role of protecting flood occurrence and prevailing of drought condition. Forest provides a cover that cushions the impact of rainfall on the hill slopes, thus facilitating infiltration of water into the soil. As a result cold transparent water is releasing through perennial fountains and springs as a main source of a river water. However, the present situation in the State is a precarious condition. Just after the rainy days, all rainwater were free to flow in the form of surface runoff, evaporation and evapo-transpiration. As a result of this, another threatening effect of man's evergrowing demand for food, shelter and drinking water is the problems of drought in the State. Now-a-days the late arrival of monsoon causes drought conditions in the valley areas particularly at the time of cultivation of crops. This phenomenon has become a regular feature in the State.

The problem in the area may be reflected as a scarcity of water. Thirty/Forty years back most of the tribal villages were located at the highest point of the hills and mountains. At present those highlanders are compelled to migrate downward where the perennial water point is available. Earlier, when the hill ranges were thickly claded with green vegetation the water sources were available even upto the highest peak. Now, the situation is no longer there.

Our Responsibility

Forests of Manipur suffer because of excessive deforestation by Jhummias and other exploiters, thus posing a threat to the eco-system. During the last 2 or 3 decades we have seen a lot of changes in the environment of the State. All the surrounding hilly ranges were deforested and left barren while the lakes and swamps were dried up. Alternately flood and drought are occurred in the same year as a regular phnomena. Landslides are the common features in all the National and State highways of the State. Still then most people are not aware of the multifacet nature of the problem. We don't understand that it is not mere protection of few animal or plant species or planting of trees, or the shutting of dams and big polluting industries or the simple cleaning up of the environment that constitutes environmental protection. It is a combination of all these. It is like the proverbial case of the elephant and the blind men - one cannot understand it unless one looks at it in its totality.

Of immediate concern to anyone interested in protecting the environment is the biosphere - the delicate layer of soil, water and air upon the surface of the earth within which it forms a part.

In order to maintain a better eco-system and a better environment in a state like Manipur, it is necessary to maintain at least 33% of the State's total area with forest coverage. The crucial question is how to ensure the maintenance of this forest area ? Some scholars argue that it can be done if jhum cultivation is totally stopped. But then, how to stop jhumming ? Most of the hills people in Manipur have historically practised jhumming. To change the age-old practice without proper and adequate occupational rehabilitation would be a suicidal proposition. Most of the policy matters and planning experts for this region have bypassed this basic question. Hence, the economic life of the hills people (or for that matter even the valley commoners) has drastically deteriorated and the forest and the eco-systems of the valley have been degraded at still a faster pace.

In a sense, it is clear that the rate of deforestation in an area is the direct impact of lower income and a marginalised existence of the forest dwellers of the State. The economic life of the Manipur hills people is marked by the ruggedness and remoteness of their inhabitations. The modern transport and markets have yet to reach their villages. Thus, they are caught in a vicious circle of poverty and environmental degradation.

In all, most of the environmental problems in Manipur, as in most other parts of our globe, seem to have their origin in the developmental programmes which are in fact purely economic in nature driven by market forces. Most of the projects in power and energy and industrial sectors directly threaten environment and ecology everywhere. The real solutions lie in the actions of individuals and communities all over the world. Corrective actions and initiatives taken at individual, group, community and village levels can only arrest the problems of environmental degradation and destruction. We may possibly suggest a meaningful slogan here : THINK GLOBALLY ACT LOCALLY. Only local, meaningful actions, not hollow rhetoric, can face the threat of environmental disaster. There is a saying amongst the Red Indians of the so-called the wild west to maintain the biosphere in a congenial balance : "The natural resources we enjoy today are not the gift of our ancestors, but we loaned these from our future generations. We have to repay it back in its natural/ original shape."

7

Determining Research Priorities in Manipur

A systematic investigation to establish a fact or cause or principles or to collect information on a subject or problem is called Research. In order to determine the priorities of Research in Manipur one has to see the relevancy of the present day economic standard along with the socio-cultural circumstances and also along with the environmental problems of the habitant. All of us are well acquainted that the state of Manipur has been physiographically grouped into three distinct zones viz., (i) The Hills; (ii) The Plains and (iii) The Foothill zones. Comparatively it is a small and remotest and communicationally backward zone of the country. However, it has been the home of the Meiteis, 29 subtribes belonging to Kuki and Naga groups and a few other communities of the latter. The growth of population in the state has been continuously and alarmingly increasing on the limited land resources of the state. One of the most serious problems faced by the state is the rapidly increasing pressure of population on physical resources, particularly on resources of land. The land area is fixed and inextensible while the population is increasing at a rate never before reached in the history of the state. With the increase of population, the per capita share of arable land decreases and consequently Jhum cycle has been reduced to 2/3 years from 30/40 years before 50 years ago in the hills. Consequently the vast forest area of the state has been reduced to a microscopic zone - causing high rate of soil erosion and soil wash in the hill slopes while rapid siltation to the river beds and drying of lakes and wetlands causing flash floods and drought in the valley.

Despite hard labours, the total economic growth in the state is much below the poverty line. Consequently, a large section of the population exist at a very low level of living. Further, it is also a fact that by 2000 A.D. Manipur will have 7.42 lakhs of unemployed. The state can provide for just 31 thousand posts only. In a state where the population pressure is checklessly increasing upon an extremely limited land resources, endless expansion of deforestation, adverse environmental changing, unplanned growth of settlement with less vision of economic exploration is never likely to be a peaceful and progressive state. Hence, this may stimulate researchers on the problems of fighting the onslaught of the State's Development and progress in different branches of knowledge which directly concern with the people of this state. The need for such researchers becomes all the more imperative in the region like ours where the available land resources is limited, the productivity is low and more than 70% of the total population is directly dependent on agriculture.

The important holistic aspect of the state's problem is the determination of appropriate man-land ratio that would fit in with the optimum production of adequate foodstuff and other infrastructural requirements. Therefore, priority of the research area should be given to the studies of Land use and its associated problems under the prevailing climatic, physiography and food habits of the people.

Secondly, researchers are required to think of a change in the cropping or farming pattern of the state particularly in the hills of Manipur. However, any change should have to be economically and environmentally viable and should also be acceptable by the habitant of this area. In this case, one may choose the problem of introduction and implementation of "Natural Farming" technique in the state.

Thirdly, most of the hill areas commonly practice the traditional technique of farming with certain ethnic law of land-use. But due to changing circumstances the system has been collapsed. Therefore, sociologically one may think on the issue of how could ethnic Law of the Land use be implemented with the changing circumstances.

Fourthly, the impact of the environmental upheavals or uncertain on the social context should also be emphasised. The detrimental effects of the Ithai Barrage, the deterioration of the fresh water body of Loktak lake and their effects on the life style, occupation as well as life-line of the people around the area pose a great threat to the very existence of the valley systems of the state. The ecological hazards to the people consequent to the construction of Tipaimukh High Dam cannot be overlooked. The social inviability and occupational loss of about 5000 people to be rendered homeless due to the inundation of 300 sq.km. of land area due to the construction of the High Dam should be given priority for investigation.

Lastly, the State Government has been taking up specific economic programme for both hill and valley. However, specific economic planning of the foothill area is yet to be done. Therefore, research priority should also be given to the foothill areas of the state.

Considering the immediate, concerned problems of the State, the researchers in the field of Social Sciences shall have to look the above mentioned problem area as priority one. Any research in these area shall have social and economic creditability. In a country like India where democratic form of government is existing, each and every individual citizen have their own right and duty to contribute for the cause of the development of their family, locality, society, state and Nation as well. *Therefore, research* for research's sake and degree only will not make a *self reliant country*. Every individual research scholar shall have the guts to combat with the problems of upliftment of the Society and Nation. Accordingly, funding agencies in the field of Social Sciences Research shall have to look into the above mentioned priority areas for funding. Only when the funding agencies implement as per the priority fixed in the context of the socio-economical and environmental relevancy of the inhabitant the scholar will effectively share their contribution in making a self reliant Nation.

8

Problems of Refugees in North-East India

The state of Manipur located in between the graticules of 23.83^{0}N to 25.68^{0}N and 93.03^{0}E to 94.78^{0}E on the globe stretching an area of about 22,356 sq.km. The area has been bordering with the following states and country : namely, Nagaland in the north : Burma (Myanmar) in the east and south-east; Mizoram in the south and south-west; and Cachar District of Assam in the west. The state has a total boundary line of about 854 kilometres in length of which 353 kilometres in length falls under the jurisdiction of International boundary line in between Burma and India. Since time immemorial the area has been settled by groups of people belonging to tribes of South-East Asian origin. Presently they are the Meiteis and other sub-tribes of Naga and Kuki groups. Now they are named as Manipuri in the present century, linguistically affiliated to Tibeto-Burman and Kuki-Chin groups of language.

Geo-strategically, the state along with other sister states of North Eastern India are all susceptible to all socio-economic problems and political upheavals happening both in mainland-India and in the neighbouring South-East Asian nations.

Since the days of Indian Independence (1947) some portions of North-Eastern India, particularly Nagaland have been protesting to the Union of India and fighting for Free and Independent Nagaland. Till now the Naga movement has been continuing in the state and further spreading to the neighbouring states of Manipur, Arunachal etc. Currently, the whole of North-Eastern states are under the activities and attitudes of separatist ideology. In the sixties Mizo movements were suppressed by Indian Forces. During the operation many civilians particularly the Hmars of Mizoram were displaced from their own state and entered into the territory of Manipur. These are some of the issues of inter-state refugees in Manipur.

At the time of partition of India and Pakistan many Bengalis, East Pakistani and Meitei refugees entered into the territory of Tripura. Assam and some of them came as far as Manipur and settled in Serou, Saiton of Manipur (1947-1950).

During 1962-67, General Ne Win of Burma while enforcing every Burmese to hold a *national identity* card drove out a large chunk of human population not belonging to Burmese. Under this dispensation thousands of Punjabees, Nepalis and Tamils entered into the territory of Manipur. They then settled mainly in Imphal and Moreh Town of Chandel district. More than thirty thousand population of Naga and Kuki groups known as CHAIN by the Burmese were also deported from Burma and settled in Chandel district, Biyang and Tipaimukh areas of Churachandpur district, Imphal and Sadar Hill areas of the state of Manipur.

With the uprising of *pro-democracy movement* during 1980 the Military Junta of Burma constituted the *State Law and Order Restoration Committee* (SLORC) to rule the country. Particularly on the Day 8-8-88 the SLORC oppressed its citizens so brutally and killed thousands of students and common people in Burma. Thousands of youths and women were used as forced labour for all kinds of hard work and military transport. Over and above, many villages like Tamu, Phailen, Tanen , Wakshu, Chumpender, Zaitiei, Nongkang etc., were forced to shift their villages to *Regrouping centres* arranged by the Military. Some of the Christian villages were not allowed to worship in some of the Kabaw valley area. As a result of these oppression, suppression and tortures many tribes belonging to Naga and Kuki fled Burma and settled in the Chandel district of Manipur. After the incident of 8-8-88 there were 120 Burmese students, supporters of pro-Democracy Movement from Arakan, Shan State, Karen, Rangoon, Mandalay, Chin Hill entered in the state of Manipur (as on 1989). They were all in the refugee camp of Leikhul.

During 1950-1980 the flow of Nepali refugee was head-counted around 40,000 population. This has been identified by the All Manipur Students' Union (AMSU) during their foreigners agitation. The flow of Nepali has since been protected under the Indo-Nepal Treaty of 1950. But such kind of free flow of Nepali refugees from Nepal has added a new dimension to the then existing problem in the hills of Manipur, resulting in environmental degradation.

After inception of Bangladesh and signing of Indo-Bangladesh Treaty of 1972 a large and continuous flow of refugees have come into Manipur. They are mostly of Muslim origin and settled in the area of Jiribam. Although recorded neither as foreigner nor as refugee, their names are found in the electoral roll of the Assembly and parliamentary elections. Almost all the refugees left Bangladesh and Nepal in search of greener pastures. Therefore, more and more Nepalis and Bangladeshi refugees entered into the territory of Manipur. Refugees from these two directions have a serious threat to the local people. They could have adopted to the terrain conditions of Manipur even better than the indigeneous people of these terrain and since then they have influenced the culture of the area. The land-use practices of these refugees have affected the normal ecological balance of the state. Further, it is also observed that because of the common racial and ethnic affinities most of the refugees originated from Bangladesh, Nepal and from Kabaw valley have somehow or other assimilated with original settlers. They have thus deprived the indigenous people of their share of product. Since there is no proper census of the number of refugees in the state, planning a realistic planning has not been possible based on local population. In a sense refugee population are able to do hard works since they are well exposed to the survival struggle and at a cheaper wage than the locals. In the long run refugee will form a major constituent in the labour and production front. And ultimately they will usurp the right of the indigenous population. Here I may be permitted to mention a Manipuri proverb "*Yumdangbạna Keinaba*" literal meaning is : refugee becomes controller of the store house. Therefore, it is high time that we look into the matter seriously and may frame a systematic - working pattern which should not affect the local people's economy.

On the contrary those Burmese student refugees who are recorded properly survive by paying Rs. 2/- (now Rs. 4) per day as pocket money. However they are provided some quantity of rice, potato, dal and Kerosene oil etc. They are confined in a cell. Sometimes, some officials utilised them as their unpaid labour. Being students, they have no chance to continue their education. They are not given proper medical treatment and recreational facilities. Under these circumstances we may call for the question of extension of human rights and minimum comfortable environment to them. These are the groups of refugees who really do not like to settle in India. Other refugees like the tribes of Kabaw Valley, Nepalis, Bangladeshis are all foreigners who wanted to settle permanently.

Inside the state, very recently, based on communal and ethnic crisis there arouse problems of *ethnic refugee*. Lives and properties of certain communities like Kuki and Naga were totally lost. Their villages were uprooted and burnt down. They are taking shelter under re–grouping centres and camps. For such an ethnic crisis and inter district refugee problems the Nation needs a specific policy which is applicable to all sections of population.

Besides these *trans–border* and *inter state problems* of refugee there are other refugee problems created by the construction of *big dam* and *multipurpose* development projects. These are the group of people who are displaced by the implementation of so–called developmental projects and dams. Their age old traditional occupation are also totally sealed. These group of people may be termed as *Environmental Refugee*. They are the indigenous people of the area where the development projects altered the natural environment totally. Under this category of refugee we may treat the main population who are totally dependant on Loktak Lake and other dams. A large number of fishing community of Thanga and Karang islands of Loktak and the Lake dwelling population of the peripheral villages were deprived of their *right to environment and occupation* in the area. For these groups of environmental refugee there is no clear cut policy for rehabilitation framed by the state or by the Centre. Even before we could settle the problem of affected people, a Dam of similar nature and more disastrous one has come up along the course of the Barak river at Tipaimukh confluence. This *Tipaimukh High Dam* will displace more than 40,000 indigenous population from the districts of Tamenglong and Churachandpur of Manipur.

The above mentioned issues are the leading refugee problems in the state. Hence,we may think on the line of having a comprehensive mechanism for dealing with the problems of refugees both Trans-border, inter-state and within the state. A twin *National Refugee Rehabilitation Policy* may also be adopted in the larger interest of the Nation and from the angle of human rights. The policy should also address the issues of the coming descendants of the present refugee generation. Every state of India should have a *monitoring cell* under the sponsorship of the Central Government, so that the refugee problems may be tackled in a scientific way.

9

Dzuko Valley – A Biosphere Reserve : An Environmental Appraisal

Introduction

DZUKO Valley and its surrounding hill ranges is a unique eco-system in the region. It is one of the rarest geographical gifts to this part of the Globe and more specially for the North Eastern India. The area has immense potentialities and hence the DZUKO region needs to be conserved as a zone of Bio-diversities in this region. Harbouring rare species, it is potentially of great genetic importance.

DZUKO Valley is located in between 25^{0}32'3" N to 25^{0}33'47"N Latitudes and 94^{0}3'43"E to 94^{0}4'28" E Longitudes. It has an area of about 2.5 sq.km. (excluding the surrounding areas). It is located at an elevation of about 2450 m. above mean sea level. It lies on the NW lap of the Essu mountain which is the highest peak of Manipur having an altitude of about 2995 m. above mean sea level.

The word DZUKO is derived from the Mao language i.e. *DZU* means water and KO means cold or frozen, hence literally called *'frozen water'*. Whole of the river surface water upto 1" thickness is frozen hard during the last part of December and the beginning of January. The valley is the repository of rugged history of the Mao-Pokmai tribes of Manipur, their religious thought, cultural resilience and spiritual energy.

In some of the Mao folk tales, the valley area were repeatedly explored in the distant past by the Mao tribes for their settlement. But because of severe cold climate and unsuccessful cultivation of crops, they had to abandon the attempt. However, the peak of the *Khongho*, the *Essu* and the DZUKO Valley had a tremendous impact on the psyche of the people here, who have touched these surrounding mountains and the DZUKO Valley with magic of their valour and love. They made the valley a natural, hollowed heritage of the people. All the time, following the changing seasonal rhythms, people, particularly youths, of *Mao* have been tempted for trekking the valley rapidly. This repeated exercise is proving of their physical fitness and mixing of morals with the nature of DZUKO. It is also reported that when the surface water of DZUKO River is frozen (Dec.–Jan) youths enjoy walking over the frozen icy surface of the river. It is also strongly believed that if a person is immoral, his or her legs would be pulled down below the frozen surface of the river.

During the 1980s, Manipur Government mooted a proposal for conserving the valley as *National Park* while the Nagaland Government worked for construction of a DAM and the creation of a *Reservoir* for solving the demand of ever increasing drinking water problems of Kohima town. Knowing all these plans the Manipur Cultural Integration Conference and the members of the MASS (Manipur Association For Science and Society) initiated for trekking the whole of the *Khongho* and the *Essu peak* to reach the silent valley of DZUKO. The first trip was made with 19 youths from Manipur valley and another 19 youths from *Mao* during 1990. The valley team consisted of Geographers, Geologists, Statisticians, Ecologists, Botanists, Civil Engineers, Artists, female and male members of AMSU and Journalists to assess the first hand information. The team travelled through the ruggest mountain bridle path of about 18 to 19 kilometers in length. The journey was well documented photographically as well as videotaped.

Physiography

Geographically the topography of the valley is an interesting one. It provides many scopes for further research in the field of Geographical features, possibilities of *glaciation* in the recent past geological era, *Rock* types, *drainage* and *slope* analysis etc. Geomorphologically, symptoms of *allochthonous* and *autochthonous* remnants were exposed in the midst of the valley portion of Manipur and along the foothills of Nagaland side. These isolated hillocks within the valley as well as the hill slopes on both sides of the valley show highly smoothened *hummocky* features. This may be one of the indications of a *glaciation* period experienced by the valley not very long in *geological past* (S. Ibotombi, 1991). Further, it has also been observed that the valley seems to have a two-tier structure - a 'V' shaped valley followed by a 'U' - shaped valley. In spite of the limitations in carrying capacity of the stream two boulders of approximately 2m X 1.5 m and 1.5 m X 1m sizes are also found on the eastern side of the bank. This is unlikely to be carried down by the stream because it has only cobble size materials in its present base (S. Ibotombi, 1991). These two boulders are presumed to be the *lateral moraines* deposited during the *glaciation period.* On the other hand the *two-tier* valley structure is not recognised in the small tributaries. There are a number of conglomerated holes, either *potholes* or *sinkholes* developed on the surface of the eastern bank near the foothill valley which may be the product of melting ice/snow or may be termed as *frost action* during the winter season. However, whether the valley has undergone a recent glaciation or not needs further geomorphic investigations before a final statement is made.

Geologically the valley lies well within the Barail group of rocks of Tertiary succession. The Barail Range lies on the NNW side of this valley comprising *Phepu* and *Japvo* peaks of Nagaland. The *litho-character* is of medium grain size sandstone having brownish gray colour. But the medium to thickly bedded sandstone in the valley has little or no intercalations of shale which is a general character of the Barails series (S. Ibotombi, 1991). In some parts of the valley well cross-bedded sandstones are also found, each crossbed having thickness of about 20 to 40 cm. The sandstones are of Quartz, Feldspars and other minerals and litho fragments. The valley itself develops along a N–S trending fault having a throw of about 25m. All the tributaries of the DZUKO river are also following either faults or joints transverse to this N–S trending fault. It may also be noted that the DZUKO valley develops on a fault plane instead of an antiformal structure (S. Ibotombi, 1991).

Soil

The soil found in the valley is dark in colour and spongy in character which may be due to its organic contents or due to its genetic factors that need further investigation. However, some sample of the soil were analysed by (Late) Dr. H Bikramjit Singh in the Ecological lab of Dr. B. Manihar Sharma, Deptt. of Life

Sciences, Manipur University. The results of the analysis are given below :-

Soil porosity	:	76.677%
Soil moisture	:	42.0%
Organic Carbon	:	5.7%
Bulk density	:	0.622 gms/cc
PH	:	5.0
Organic matter	:	9.826%

Climate

It appears that the DZUKO valley region enjoys *cold temperate climate* during the winter season while *heavy rainfall* during monsoon and *pleasant temperature* during the summer. However, the region is yet to be measured in terms of the mean annual rainfall, temperature and humidity etc. It is presumed that the area has an average humidity of about 60% with a maximum of 85 to 90%. Likewise the average temperature of the region has been estimated at minus 6°C below zero to 18°C. Snow falls during winter is common. During the end of December and the beginning of January trekkers have come across with snow capped portion on the Northern slope of the Essu peak, the frozen streams and DZUKO rivers.

Forest Types and Flora

As per revised classification of Indian forest types by Champion and Seth, 1968, the forests of the area belong to Montane Temperate Forests under Group II–Montane Wet Temperate Forests, Sub-Group IIB-Northern Montane Wet Temperate Forest in C1, East Himalayan Wet Temperate Forests and C2 - Naga Hills Wet Temeperate Forests (Type IIB/C1 & C2).

Flora

The team visited the region every year from 1990 to 1995 in different seasons for proper documentation and study of the area. The same team took two different routes viz. (1) Mao IB Via Potato Farm and the Khongho peak, (2) Mao IB Via T B Sanitorium and the Khongho peak. The first route is about 18 to 19 kms. while the later one is about 9 to 10 Kms. in length. Both the routes meet at Khongho peak (2700 m.a.m.s.l). During the trip in between Mao IB to Khongho peak, the team observed broad leaved trees, shrubs, herbs and different ephiphytes and climbers. The broad leaved trees are *Alcimandra catheartii, Schima wallichii, Rhododendron arbarium R. elliottii, Fatsia* sp. etc. Among the shrubs, *Lyonia ovalifalia, Osbeckia, Mussaenda* etc. were recorded. Herbals like *Anemone, Calanthe* and several others also grow luxuriantly. *Coelogyne, Bulbophyllum, Dendrobium, Aeschynanthus* are among the *epiphytes. Glaeichenia, Leucostegia, Davallia, Asplenium, Polypodium, Drymaria, Drymoglossium, Lycopodium* and different species of *Selaginella* are common under the pteridophytes (Ref. Late Dr. H. Brikramjit, 1991).

In between the *Khongho peak* to *Mt. Essu,* the team came across a vast track of *Tenwa Forest (Arundinaria)* with mixed stunted plants of different species. In between North Western slopes of Mt. Essu and DZUKO Valley, the team observed *Rhododendron manipurensis, R. macabeanum* growing with exorbitant flowering. The DZUKO Valley hill slopes and foothills were covered gradually by big trees of timber varieties (Ref. Late Dr. H. Bikramjit, 1991).

From distant hill ranges the valleys appear fully carpeted by green spongy materials dotted with different colouring designs of seasonal flowers. During November to March, the valley floor looks grayish yellow predominated by the dart bamboo (*Arundinaria*). All the streams, tributaries and the main DZUKO river were frozen during most of the days in December and January. During June to July, the valley was fully bloomed with purple colour flower of *DZUKO Lily (Lilium chitrangade – Bikramjit).* Many yellow flowers of *Ranunculacease* and *Scrophulariceae,* purple colour flower of *Kaempferia* (Zingiberaceae), were also visible. During July to November many varieties of white flowers and dwarf plants covering the whole of the valley and the surrounding foothills were observed. The team also recorded a few species of *terrestrial orchids* in the valley. During the trips (Late) Dr. H. Bikramjit collected more than twenty species of ferns, forty *Dicotyledonous* plants and about thirty *Monocotyledonous* plants for identification. The best flowering season of the valley is June to July when the vale glitters with the blooming of several species of plants. During the dry winter also a few species could be seen in flowering.

Fauna

Due to the rich flora of the valley and its surroundings and the lack of human accessibility, different species of birds and animals have found a *Safer home* in DZUKO Valley. However, during our subsequent expeditions we happened to see some unidentified *fowls* and other *birds*, pairs of *Cobra* (medium size) and also some of ferocious *wild cats* etc. There are records and accounts of Elephants being regular visitors to DZUKO Valley (E.W. Dunn : 1873). However, in some of the documents, it is reported that the following *mammals* and *birds* were seen. They are namely *Artelop Sp. (L), Himalayan Black Bear, Lesser Cats, Sambar Deer, Barking Deer, Wild Dog, Flying Fox, Hoolock Gibbon, Langoor,* etc. (S. Seth : 1991). Various insects and invertebrates other than insects inhabiting the valley were observed. Among them a number of exotic *butterflies* and *earthworms* may be mentioned. Among the birds, the Great Himalayan Barbet, Eagles, Green pigeon etc. were also observed (S. Seth : 1991). Among the *amphibians*, three rare species of high altitude frogs were seen in the Valley. They are, namely, *Hyla anaetens, Rana verrucosa* and *Megophyrs monticola* (Dr. Kiesetto, 1991). The above mentioned faunal species are not meant to be exhaustive and there are numerous species of both flora and fauna that need to be identified.

Development and Dzuko Environs

The Nagaland Government has proposed many developmental projects as mentioned below :-

1. Paper mill using *Tenwa (Arundinaria)*
2. Hydel project
3. Sheep farm
4. Mithun grazing field
5. Luxury Tourism with 300 huts
6. Boating with the artificial flooding of the valley etc.

Currently the Nagaland Government has concentrated its efforts for the implementation of the following plans viz.,

Water Supply Scheme to Kohima

Hydel Project and Tourism

Accordingly, the Govt. of Nagaland has constructed two different roads. They are, namely, *Jakhama DZUKO Road* (7 kms.) and *Viswema DZUKO Road* (13 kms.). All these roads were constructed upto the length bracketed above, but these roads could not be extended further because of the lack of gradient crossing a huge bend of sandstone Hilly range. On the other hand, road construction has been banned by the Naga sensitive people and student organizations who are working for the protection of *DZUKO environment* and its *Natural ecosystem*. They further objected to all the proposals of the Govt. (developmental plans) stated above. Further it is also observed that the feasibility report on water supply project submitted by WAPCOS [Water and Power Constructive Services (India) Limited] does not reflect on environmental impact Assessment of the DZUKO and surrounding areas after inundation.

Although we need continous scientific environmental impact study in the area, however, a number of environmental problems can be foreseen if road construction and developmental projects are implemented.

The DZUKO area is one of the most interesting *genepool* of *endemic bio-diversities* of this region. To protect the area is the service to all human kind of the North Eastern India particularly of Manipur and Nagaland. There should not be any territorial dispute in between Nagaland and Manipur for DZUKO Valley. The valley is the gift of Nature to all the mankind of this region. Hence, the academicians, environmental activists and all other people should fight together to save the *DZUKO Valley* from the claws of the so-called developmental projects. The people of both sides should decide by sitting together to declare the *DZUKO Region* as one of the rarest *Biosphere zones* and *common heritage* of Manipur and Nagaland.

Conclusion

In conclusion it may conveniently be inferred that DZUKO - a Nature's conservation - Unit must be preserved as it is. And this must be the nature's labs to promote intensive and integrated research activities covering most of the disciplines specifically concerned with the area. This will be possible only when both the Governments of Manipur and Nagaland take special interest and initiative in collaboration with the associated departments of Government of India, other Global bodies and the voluntary organisations specially of Manipur and Nagaland.

10

Ithai Barrage : A Threat to Natural Drainage System of Manipur Valley

The valley of Manipur, according to literature and legends, historical evidence as well as scientific studies, was formed in course of time from its fluvio-lucustrine origin. During its formation the topographic system also developed its own, natural unique drainage system, with underlying rocks throughout the length and breadth of the valley . Any so called ' modern', technological, unnatural intervention into this system, without proper and deeper understanding and planning, is likely to pose a threat to the natural compact drainage system of the valley. Therefore it is of immediate necessity to trace the chain reactions caused or likely to be caused by the construction of Ithai Barrage on the southernmost course of the Manipur river. A detailed scientific and diagnostic analysis of the negative and positive effects of the Ithai Barrage on the natural grainage system of the valley is a must at this stage when the valley and the people inhabiting the valley are faced with the danger of a total ecological disaster. Such an analysis has to go hand in hand with the study of the rainfall behavior in the state. Additionally, it is also highly essential to study the threat of rapid increase in population, the change in demographic composition and changing patterns of land use not simply in the catchment areas of the Manipur river drainage system and related natural phenomena of the entire valley itself.

Geologically, it is believed that the valley was formed 30 million years ago during the Tertiary era. Further research and investigation have proved that the valley formed itself as the result of the tectonic disturbances in the course of the Manipur river near Serou. This area is popularly known as Sugnu Hump/ Serou Hump. Infact, the emergence of Serou Hump affected not only the normal flow of the Manipur river but also contributed to a development of a delicate balance between the normal rate of erosion of the river and depth of the river bed. It is precisely for this reason that this portion of Manipur river is termed as 'antecedent drainage' by the geomorphologists. From the common sense understanding, the flow of the Manipur river declined near the Sugnu Hump as the level of the river bed also rose up. Consequently during the monsoon, an immense amount of water used to get accumulated and impounded upward to form a vast sheet of water. However, in course of time, the level of water receded and formed the habitable valley with a number of lakes scattered all over the valley, with, of course, major concentration in the southern part of the present valley. Many of the big lakes like Loktak Pat and Pumlen Pat etc. are alive still at the low lying belt of the southern valley.

At present, the drainage system in the Manipur valley is a natural gift of the system and network of the Manipur River consisting the major rivers (its tributaries) viz., the Iril, the Imphal, the Thoubal, the Sekmaijin, the Khuga and the Chakpi. In the interfluve areas of the Imphal and the Iril rivers, Manipur river and Thoubal river and Manipur river and Sekmaijin and the Chakpi, there were many low lying areas naturally forming lakes and shallow lakes supporting the regime of the Manipur River. These lakes and shallow lakes of the interfluve areas served as natural reservoirs for the excess monsoon water of the respective rivers. This also played a very positive role in the sustainable development of the micro - eco - systems of the localised areas as an integral part of the entire eco - system of the Manipur valley. These in turn become excellent breeding grounds for the rich flora and fauna of the valley and particularly for the indigenous fishes of the State. For example, the Loktak lake, one of the biggest natural fresh water lakes in Asia, is the major reservoir of the Manipur River. During monsoon, all the rain water from the different catchment areas of the tributaries of Manipur River got collected at the Manipur River and then naturally drained towards the southern side of the valley, where the Loktak, the Pumlen and the Lamjaokhong lakes have been situated. Unbelievably, the rain water which used to get impounded by the Sugnu Hump and also the torrential obsiquent stream of the Chakpi river receded upwards upto the point of Khordak channel of the Manipur river where the Khordak channel allowed its inflow into the Loktak. In short, under such organic and natural conditions the Loktak lake acts as a reservoir for the excessive water of the Manipur River. Still more, the Khordak channel allowed the drainage of the water from the Loktak to the Manipur river during the lean seasons through Khordak itself. Thus, Khordak acted as both the inlet and outlet for the maintenance of a delicately balanced level of water. And hence, the Loktak lake itself is a part of the Manipur river drainage system.

The marvel of nature does not end there. The forest coverage in the state, of the quantum of rainfall received in the catchment areas of Manipur river system and the pattern of land-use practices within the catchment areas all related to each other and formed an organic link between the existence of the lakes and also the occurrence of flood and drought in the valley. The untampered laws of nature symbolised the coexistence and survival of the people of the hills and plains in Manipur unless we understand the history and the laws and linkages of nature gifted to our state, we might disturb this coexistence by intervening unwisely and violating the entire system handed down to us as gifts of nature.

To come to more details, the equilibrium maintained between the forest cover (at least 33% of the total area of the state), land mass and the density of population, from the dawn of history, has helped our state received optimal rainfall for optimal productions in agriculture, and the prevalence of pollution free, healthy and pleasant climatic conditions. This short of an equilibrium, an intricate balance between man and nature has been observed to exist upto about 1950 in Manipur.

It is obvious that the rapid and continuous growth of population within a limited land area would cause pressure on the forest cover. This is what has tragically happened to Manipur. Around 800 km^2 of forest cover is being cleared by the Jhummias for shifting cultivation and by other people for firewood and timber per annum. Continuous pressure on land naturally reduces the 'Jhum Cycle' from 30 - 40 years (1950) to 2-3 years as is at present. This has been clearly shown through remote sensing photograph interpretation in 1990. As per the Report on Land use/ Land cover, Manipur State of SRSC, Deptt. Of STE, Govt. of Manipur and HRSA, Deptt. Of Space, Govt. of India (January

1990) only 3% area of Manipur State is under effective forest cover while rest of the forest area belongs to degraded/deforested area. Over and above all this, the wave of higher degrees of consumerism and commercial exploitation of forest violate all natural laws and threatens the eco - system of the State. Consequently, more than 90% of the hilly terrain has become barren. Consumerism and commercial exploitations have also increased the siltation rate tremendously along the river beds and lakes in the valley. Even rare and endangered plant and animal species have not been spared and face a grave danger of the highest degree.

Things have become worse to worst because of the ill - conceived planning of projects for Manipur, in the pretext of so - called 'Development'. The Ithai Barrage, for power generation, has been constructed at an unstrategic and unscientific point of the Manipur River, i.e. after the confluence of the Manipur and the Khuga rivers so as to devastate the future of the future generations of Manipur. To recapitulate, 2/3 decades ago the common people settled in and around the peripheral areas of Loktak, hoping that the completion of the Loktak Multipurpose Project (LMP) would boost their living standard and economic conditions. Unfortunately, the opposite was in store for these innocent people. Even the three crucial promises of the project have proved false :

1. Instead of reclaiming 40,000 hectares of surplus land from the peripheral areas of the lake and its distribution to the displaced/affected people, more than 30,000 hectares of land has been getting submerged after every shower. This has deprived a section of the peasant population of the region of their right to land and life.

2. It was promised that the LMP will provide continuous supply of water by lift irrigation for multiple cropping (60,000 hectares of land) in order to achieve adequate production of food. However, the LMP itself is starving of water. For lack of water, the power generation capacity of LMP has gone down to such a level that the power production unit had to be closed down in 1992.

3. It was also promised that the electricity produced by the LMP would be sold to the consumers in Manipur at a rate below

10 paise per unit. But presently the people pay about one rupee per unit. Most of the people innocently believed that after the completion of LMP, not only there will be proper industrial and agricultural growth in the State, but also the project Trolly network would be introduced to provide pollution free transport facility throughout the State. Unfortunately, the people, on the whole, have been cheated by the LMP beyond imagination.

Anyone who studies seriously the Project Report of Ithai Barrage with a proposal for the construction of a dam at a height of 760 metres contour line will be utterly shocked. The obvious implication of a 760 m contour line dam is that the entire surrounding area would be submerged under water and the people inhabiting the area would be displaced, leading to their marginalisation and pauperisation. If we analyse the toposheets, we would arrive at a scientific and convincing conclusion that the water would reach upto Hiyangthang including many other settlement areas of Moirang town itself. That is why a large tract of area has been flooded with water more frequently than ever.

Now, after the construction of Ithai Barrage, the entire eco - system of lakes has terribly deteriorated. All the fishermen communities and lake dwellers have suddenly found themselves totally uprooted. Driven to the wall, the fishermen have recently resorted to application of pesticides to catch fishes. This has in turn led to the pollution of lake water and health hazards for the fish eaters of the State. All these implications were pointed out well in time to the Project authorities i.e. in 1975 by a group of concerned people led by M.K.P.B. Singh. But it all fell into deaf ears.

All the devastating consequences for the men and nature of the State noticed today indicate the irresponsibility of the planners of the project. In fact, no systematic survey before the construction of the project was over done. Even no environmental impact assessment was done. Naturally, the ill - conceived implementation of projects has resulted in all round destruction rather than development. The rate of siltation in the lake itself has risen to an unmanageable extent. And no serious attempt is being made to control the siltation rate by protecting the catchment areas of the Loktak bound rivers or by any other scientific method.

Today, the people of the villages directly affected by the Ithai-Barrage has become volatile about the grave danger to their existence. We can only hope that everything is not lost. We might initiate discussion on certain issues which might lead us to reduce the situation at least practically and this must be done immediately and urgently within the framework of natural laws and the organic relationship between men and nature. Let me try to put forward some suggestions. Since the Ithai Barrage was constructed at the lowest point after the tri - confluence of the Manipur River and the Khurga and the Ungamel, the water receded both northward, rather than southward as was the natural course, to Loktak lake through the Khordak channel. According to the information of Loktak Development Authority (LDA) levelling staff itself, the actual depth of the Khordak channel bottom is hardly ONE metre whereas the main river has a depth of 6-7 metres below the inlet channel. Therefore, a maximum amount of water cannot pass into the lake and hence more and more land area would increasingly be submerged under water. Daming at Ithai has also slowed down the current of flow of water leading to greater deposit of debris into the river bed and the lake. Before the construction of Ithai Barrage, the Khordak channel functioned as both the inlet and outlet channel and hence there was no possibility of depositing much debris. As said earlier, from time immemorial, the flow of river water was controlled by the Sugnu Hump and also that the Chakpi river played the obsequent role in a natural process. But all these natural courses have now become defunct because of the Ithai Barrage, posing a severe ecological threat to the lake and surrounding areas. Again, it must also be realised that the water level of Loktak lake is higher than that of the Pumlen Lake and the lower has been blocked by daming at Ithai so water flows into the Pumlen Lake resulting in further increase in the submerged area. Thus, the entire Chairel village has come under water and the inhabitants of the village have been forced to shift to the surrounding hill slopes.

The proclaimed purpose of the Ithai Barrage was basically to maintain a particular water level at Loktak lake for efficient power generation. Before the Ithai Barrage's commission, the water level of the Loktak lake fulcated, rising and falling seasonally. Such seasonal rise and fall in the water level was a tremendous source of energy in terms of Bio - energy circulation. The fishermen communities enjoyed both the seasons - one during the rise and the other during the fall (e.g. phoom namba). During the fall in water level, most of the dry land inside the lake formed the ground for many varieties of aquatic edible vegetables. On top, the floating 'phum' in the Keibul Lamjao Wild Life Sānctuary sank during the dry season getting attached to the ground level and thus collected the nutrients from the soil so as to enrich offshoot of grasses which is essential for 'Sangai' the rare and endangered Brow antlered deer. Now all these inter - related systems have been destroyed and the destruction is irreparable.

On the other hand, because of the continuous deposit of silt and debris into the Loktak lake, the depth of the lake particularly the storing capacity of the lake has been fallen drastically. As a result more and more of the peripherial land bringing under water, deprive the people of their right to life and livelihood.

The so called 'technocrats' have been advancing an argument that the Ithai Barrage is a great means for flood control. Thus they went ahead with the destruction of natural dam of Sugnu Hump. These technocrats have bypassed all scientific or academic discussion. No examination of the unique character of the upstream and it's catchment area has been carried out. No doubt, the removal of Sugnu Hump has led to disturbances in the hydrological cycle and the humidity balance of the valley has been on the decline.

Being displaced by the LMP and Ithai Barrage and also being cheated out by the failed but false promises, individual families were pushed to go into encroach upon the whatever land that was available and into the lake causing popular resentment.

In the end, I would like to put forward some concrete suggestions : as an alternative to Ithai Barrage which has resulted in the inundation of a vast amount of agricultural land and also to the overflooding of Pumlen area.

a) Diversion of Khuga flow at a suitable place or near Kumbi to the Loktak and also diversion of the Manipur river's flow near the Nongmaikhong area to the Loktak. Thus, the submergence caused by the Ithai Barrage may be precluded and the adequate supply of water of Loktak for power generation may be maintained. As an immediate programme of action, the dredging of Khordak Maril, Yangoi Macha, Yangoi Achouba will certainly help in the proper flow of water to Loktak.

b) Afforestation programmes and protection of forest cover all over the catchment area of the major tributaries of the Manipur River will certainly regulate the flow of monsoon water and timely checking of excessive soil wash, so that silt and debris deposits along the river bed and lake can be prevented. On the whole, the alternative occurrences of flood and drought can be done away with. And also the longivity of the Loktak lake can be increased.

c) The flood water stagnation in the area lying between the Ithai down stream and the Sugnu Hump may be saved by diverting the Chakpi river, without disturbing the Sugnu Hump at all.

d) Coordination among various project authorities i.e. Imphal Barrage, Thoubal Multipurpose Project, Iril Pilot Project, Khuga Dam, Sekmaijin etc. is needed so as to regulate the flow of water and rating of water into the Loktak lake as well as provision of irrigation facilities in the State.

e) Lastly, undertaking a systematic land - use survey both in the hills and in the valley, proper study of the Manipur River network will positively control and regulate the desperate encroachment into land by the displaced people in the catchment area and on the river bank and also into the lake itself.

These are all tentative suggestions and not final. What I would honestly appeal for is to study the organic link between men and nature in a socio - historical context. Any extraneous intervention without such an understanding will not only destroy the people and nature of Manipur but also the inventions themselves. Let us save nature to save ourselves and build a beautiful future full of plants, flowers and food, fishes for our succeeding generation. Else, they will not pardon us and we will be pleading guilty to our children.

11
Tipaimukh High Dam and its Impact on Environment

Genesis of the THD

The proposed Tipaimukh High Dam (THD) has been initiated since 1929 under different names and different locational points across the river Barak. According to the 1984 Project Report, stated objectives of the Dam is an integrated approach for generating hydroelectricity, taming of floodwater and provision for irrigation etc. The first site of the Dam was located at Mainadhar (1955) and the second at Naraindhar (1964) about 40 km down stream of the Mainadhar. The third site again was proposed to be located at Bhubandhar. The fourth and present site of the Dam is located at about 500 meters downstream of the Barak after the confluence of Tuivai river. The area is popularly known as TIPAIMUKH, a border village of Manipur. Taking the name of the area, the proposed Dam is called Tipaimukh High Dam. The Dam site has a distance of about 95 km from Fulertal by rafting the upward of Barak, 334 km from Imphal via Churachandpur and 180 km from Aizawl by road.

During the first decade of the post - independence India, the planners of the Nation believed that large and high Dams are the "Temples of Modern India". However, that faith has been lost by the experiences of large dam projects which have led to a rethinking regarding their overall beneficial impacts, on the magnitude of adverse social and ecological effects. In fact, experience shows that the irrigation, flood control, electricity benefits have turned out to be not only exaggerated but negative in some cases. On the other hand, a growing people's consciousness and movement to safeguard their cultural and natural heritage against rapid destruction in the name of Development has been increasing day by day. That is why people's movement against Tehri, Sardar Sarovar, Narmada Sagar,

Suvarnarekha, Keol Karo etc. (and Loktak project) have intensified in recent years - names of all these large dam projects do not immediately conjure up the vision of progress and prosperity; rather the first memories these evoke are of the struggles against the adverse social and environmental effects of these dams. Particularly, people of Manipur are experiencing the worst adverse affects of the Loktak Multipurpose Project to its eco - system. Even today, not a single objective targeted by the Loktak project could be achieved. Despite the sad cost of the social and environmental heritage of the directly affected indigenous population living in the submergence area who have to be evicted for the creation of the reservoir, it is also likely to affect adversely several other persons for whom the impact becomes clear only much later.

Geo-Ecological Setting of THD

The total geographical area of the Upper Barak Manipur region is approximately 13,300 sq. km., lying in between latitude 23^0 10' N - 25^0 40' N and longitudes 94^0 E - 950 20' E. The catchment area of the Upper Barak Region (UBR) occupies the areas of Senapati, Tamenglong and Churachandpur Districts and Jiribam sub - division in the state. The main tributaries are Irang, Leimatak, Makru, Tuivai and Jiri. The main headstream of the Barak emerges from a village named Liyai Khullen (1890 m.a.m.s.l.) of Senapati District bordering the Ukhrul District of Manipur. The main trunk of the river initially flows westward to southward. It continues in the same trend upto the confluence of Tuivai. Then it flows northward to westward after the confluence with the Jiri river at Jiribam. The most dominant drainage pattern in the area is "trellis" type with secondary tributaries elongated parallel to the master stream (SRSS - UBRW :1997). Dendritic drainage pattern is typically developed in the northern and eastern parts of the area. The lithology and geological structures show high degree of control of the drainage pattern in the western part as compared to the area to the east.

The ridge and valley type of topography is developed in the northern and eastern part of watershed, typically in folded mountain belt with alternate soft and hard formation of rocks. The massive rocks are present in the central part of the region. There are rapids and waterfalls along the courses of the Upper Barak Region.

Geologically a series of sedimentary rocks, mainly shells and sandstone, constitute the whole of the UBR. These sedimentary rocks are subjected to tectonic activity affecting the anticlinal hills and synclinal valleys in the region. These rocks belong to Disang Group (Eocene) overlained by Barail (oligocene), Surma (Lower Miocene) and Tipam Group (Upper Miocene). The recent Alluvium (Pliestocene to Recent) is deposited along the river valleys. Further it is also observed that the whole of the Upper Barak Region drainage basin is mostly formed by faulting and fracturing of rocks. Numerous well defined lineaments trending N - S, NNW - SSE, NEE - SW, NW - SE and NNE - SSW have been observed.

As per available land use statistics the whole of Upper Barak Region may be classified as follows :-

Land Use Pattern	Area in sq. km.	P.C.
1. Thick Vegetation/Forest	2,374.21	20.00
2. Moderate vegetation/Forest	2,864.16	24.00
3. Sparse vegetation/Forest	4,302.87	36.00
4. Current Shifting Cultivation	282.96	2.41
5. Degraded Forest	1890.68	17.00
6. Alluvium	18.05	0.15
7. Water Bodies	0.62	0.01
	11,734.00	99.53

(*Source* : NRDMS, 1991)

Further, two types of agricultural practice have been observed in the UBR. The flood plains and their depositional area are being used for wet cultivation whereas the moderate to steep slopes for Jhumming cultivation. More than 60% of the UBR has been covered with bamboo, cane and mixed forest type. The region has been extensively settled by the indigenous tribes of Manipur. They are Pukmai, Zemei, Rongmei, Liangmei, Ipumei and Hmar tribes of Manipur. These are the groups of indigenous people settled all over the UBR along with their age old traditional heritage. Through centuries, these indigenous people live in complete harmony with the natural surroundings. They enjoy the forest for their livelihood and healthy culture. But after the proposal of construction of THD in the region, the normal environmental features are threatened severely.

Its Impacts on Environment

The Tipaimukh High Dam (THD) has been basically designed for controlling the frequent flood in the Cachar plains, while permanently submerging an area of 275 sq. km. in Manipur State. In this submergence, the people of at least 31 villages of Tamenglong and Churachandpur districts of Manipur will be forever displaced from their own soil. More than 40,000 people will be deprived of their Right to environment an environment in which they were born and have survived for a long period. Their indigenous occupational activities will be permanently sealed. The submerged area will be transformed into an artificial lake, with recreational centres and tourist resorts. In this process the 20 km. long area of New Cachar Road (NH 53) including two sizeable bridges will be under water. This NH 53 is the only prominently emerging lifeline of Manipur with other parts of the country. This submergence will also block the traditional water ways which link (by rafting) the remotest villages of UBR. Many weekly markets are located all along the banks of the Barak and they are linked to villages by country and diesel boats, floating of bamboo and cane etc., making a real life line in the zone. All these facilities will go down the reservoir of the Dam.

Tectonically, UBR is one of the high seismic sensitive zones in the world, with an evidence of 8+ Ritcher scale in the past. It is a well known fact that in several cases huge quantity of water in reservoirs have led to increase seismic activity in nearby areas. This evidence has also been reported from several parts of India. In 1967 a dam induced earthquakes in Kyanagar, Maharashtra has been recorded. Environmentally, not only the UBR, but the whole of the state is yet to be surveyed and documented systematically. It is necessary also to define the gene pool of the region in particular. All the habitats of the state are yet to be surveyed scientifically. If someone visits the area, (Dam site) even a casual observer will say that Hoolock Gibbon, Tortoise, Hornbills wild fowl, different varieties of reptiles and amphibians etc. are plenty in the region. During his course of research work, Dr. W.

Vishwanath Singh (LSD - MU 1983) reported more than 150 rare species of fishes in the Barak river. Some of them originate from the Bay of Bengal coming upto the head stream of the dam for laying their spawning. The construction of the dam at Tipaimukh will certainly disrupt the route of spawning areas. Further, after the construction of the dam water flow of the Barak will reduce downstream and this will adversely affect the fish population and fishing in the region. This has clearly been expressed by Y.P.Murthy, a senior engineer in his paper "Environmental problems of water - resources Development in the Himalayan region : "The construction of a dam across a river has a profound effect on the fisheries and their development." It will also have long term effects involved in changing the ecology of both downstream and the Deltic region, Therefore, it is clear that the possible adverse effects of THD will be much greater in submergence area, and unless adequate and careful attention is given to these possibilities, the Dam can easily end up causing more misery than prosperity it brings for some. Today we are faced with a situation of new - found world wide concern for social and environmental aspects of large dams co - existing with giant multistage projects which fail to incorporate and reflect those concerns in a satisfactory way. This is the situation not only in India but in several other countries as well. Humanitarian thinking based on longest experiences is miles ahead of the big projects which were planned several decades before perhaps even this information was available or if it was available or if it was available it had not been analysed and compiled in an effective way to convey the message that some re-thinking could be possible. This is the same situation in the Tipaimukh High Dam. The proposed THD is likely to affect a very large number of people and environment of UBR in a big way. In this case social activists, environmentalists and even some reputed scholars are saying that the project is not really desirable from the point of the welfare of people, especially those from weaker sections. Therefore, it is necessary to incorporate the latest information and thinking on social and ecological impact in the region. Further, it is also necessary to draw up a list of such controversial projects in the state for comprehensive re - examination over a period of, say, about two years by involving local leaders and people's organisations etc. However, there should not be any conflict between the so - called Development and Environment. The

Development process should be people - oriented and environmental friendly one. In this context, it is necessary that the planners direct their planning for development which is in tune with nature.

If some one seriously diagnoses the symptoms of developmental projects allthrough the Upper Barak Region it will be found that there are more than 39 points where micro - hydel projects can be located. If these micro - hydel projects are implemented, they would certainly help in moderating the flood water at different levels of watershed and catchment zones of the Barak and its tributaries without affecting the environment, and hence flood problems of Cachar plains can be moderated and also ensure perennial water supply in the command areas of the Barak River. By these 39 micro - hydel projects alternative to one huge Tipaimukh High Dam, we will also be able to generate a wave of equal - area development all over the entire Upper Barak Region, by extending infrastructural facilities in the area. The required amount of power which is currently needed by the state and adjoining areas can also be generated. One of the brightest example is the case of China where 75% of the power needs of the country comes from micro - hydel projects but not from big or high Dams.

12

Wetlands of Manipur

Introduction

Wetlands are areas where water is the primary factor controlling the environment and the associated plants and animal life. They occur where the water table is at or near the surface of the land or where the land is covered by shallow water. The Convention on Wetlands (also known as the Ramsar Convention) takes a very broad approach in defining wetlands. According to the convention (Article 1.1), wetlands are defined as :

> *"areas of marsh, fen, peat land or whether nature or aritificial, permanent or temporary with water that is static or flowing, fresh, brackish or salt, including areas of marine water the depth of which at low tide does not exceed six meters".*

In additon, there are human-made wetlands, such as, fish and shrimp ponds, farm ponds, irrigated agricultural lands, salt pans, gravel pits, sewage farms and canals.

Wetlands occur in every country, from the Tundra to the tropics. How much of the earth's surface is presently composed of wetlands is not known exactly. The world Convention Monitoring Center has suggested an estimated 570 million hectares (5.7 million square kilometers), roughly 6 per cent of the earth's land surface, of which 2 per cent are lakes, 30 per cent bogs, 26 per cent swamps and 15 per cent flood plains. Mangroves cover some 240,000 square kilometers of coral reefs remains worldwide.

Wetlands derive their value from their hydrologic, biologic and physico-chemical characteristics. The interaction of physical, biological and chemical components enables the wetlands to perform many vital and complicated functions. Some of these are water storage, storm protection and flood mitigation, shoreline stabilization and erosion control; groundwater recharge (movement of water from the wetlands down into the underground aquifer); groundwater discharge (the movement of water upward to become surface water in a wetlands); retention of nutrients; and stabilization of local climatic conditions, particularly rainfall and temperature.

Wetlands are among the world's most productive environments. They are cradles of bilogical diversity, providing the water and primary productivity upon which countless species of plants and animals depends for survival. They support high concentrations of birds, mammals, reptiles, amphibians, fish and invertebrates species. Of the 20,000 species of fish in the world, more than 40 per cent live in fresh water wetlands. Wetlands are also important storehouse of plant genetic material. Rice, for example, which is common wetlands plants, is the staple diet of more than half of world's population.

In addition, wetlands are important, and sometimes essential, for the health, welfare and safety of people who live in or near them. They perform a number of vital functions and provide a wide array of benefits. All wetlands values and benefits, whether for fish production, agriculture, other food production, groundwater recharge, recreation, waterfowl habitat, flood control, etc. derive from the maintenance of the fundamental characteristics of wetlands.

Wetlands of Manipur

Manipur because of its peculiar geographic situation in the eastern Himalayan offshoot region and watershed area of two great river basin systems of the world – the Brahmaputra – Barak and the Irrawady – and stradding the international border between India and Burma, is blessed with an enormously rich and diverse heritage of wetlands. Wetlands in Manipur are mostly found in the MANIPUR Valley and are known by the generic name PAT, but the surrounding hilly terrains are also replete with many kind of wetlands, great and small. In addition to the lavish natural endowment of this State, indigenous population of this area have also built up a highly sophisticated and endemic engineering marvel that extends the wetlands, and their benefits – the traditional drainage and flood control system of KHONGS, KHONGBALS and artificially developed TURELS.

The wetlands of Manipur are now known to have their own life span and hydrologic characteristics related to the evolving geo-physical character of the land itself. The probable genesis of the wetlands of Manipur seems to be caused by a tectonic movement resulting of rising the riverbed of the Manipur River at its lower course near the present Serou village. This part of the rising portion of the Manipur River is known as Sugnu Hump. Rising the river bed in this particular area makes impounding of water and forms a vast sheet of water. In due course, the water receded and formed the present Manipur valley with a large number of wetlands. Thus some of the geomorpholgoists believe that the Loktak and other sister lakes are the remnants of the vast sheet of the water and part and parcel of the drainage system of the Manipur River. Directly or indirectly, the wetlands of the southern Manipur valley supports the existence of various biodiversities of the region. It acts as catalytic agents of the whole range of natural eco-systems in the region and also allows the transmission of various wavelengths of socio-cultural and politico-economic activities in the state of Manipur. Over and above, these PATS are the main source of bio-energetic for the entire state of Manipur and consequently a large section of rural peasantry in the vicinity depends on it for their livelihood. In short people of Manipur fully exercise their 'right to life' and 'right to environment' on these wetlands of the state. Further, it is also interesting to note that in all the interfluve areas of the Manipur river drainage system; one or two wetlands are always formed by NATURE as natural heritage of the area.

Considering all these vital roles played by wetlands, we need detailed analytical studies of the existing wetlands of the state. Different parts of Manipur are now known to have their own life span and hydrologic characteristics related to the evolving geo-physical character of the land itself. Further, it is also necessary to study the changing, degraded ecosytem of the state because of effects the increasing trend of population pressure and its associated land use pattern of the state. These multivaried studies will certainly throw certain light on the effects and loss of the wetlands in the state of Manipur.

In the beginning of this century, there were nearly 500 PATS (wetlands) in the Manipur valley itself, covering a large portion of its area. A part from the Loktak, by far the most important and largest, other well-known pats of Manipur valley are Ikop, Waithou, Ngarka pat, Loushipat, Leingangpat, Sanapat, Laisoipat, Laphupat, Loukoipat, Kharungpat, Khonghampat, Lamphel pat, Porompat, Akampat, Takyelpat, Kombireipat, Yaralpat, Poiroupat, Takhokpat, Kakwapat and many others. In addition to these valley-based wetlands, there are a number of highland lakes and swamps, as well as numerous rivers and streams. It may also be mentioned that there were many ponds and dug-out tanks for each and every house of the valley before the present decade. Now almost all the ponds and lakes have been filled with earth and occupied for human settlement and commercialization of space etc. Hence, the overall hydrological status of the Manipur valley has been fast changing. In the mean time, the environmental conditions of the state has become degraded to such an extent that the area under green coverage/forested area has been greatly reduced. Consequently debris brought down from the catchment area of the valley-based rivers have been continuosly deposited into the valley wetlands. Thus almost all the existing wetlands have become shallower and shallower. Besides, the construction of dams (like Ithai Barrage) makes more and more siltation along the riverbeds. All the factors stated above together leading to occurrence of frequent flash floods and alternately, drought in the cultivating seasons are becoming a regular phenomenon in the Manipur valley. As for example, the negative consequences after the commission of the Ithai Barrage and continuous degradation of Loktak environment, including Eutrophication of lake, are well known by all sections of population in this region. Further, extension of surface communication network inside the Loktak lake areas, for example, construction of surface road in between Bengul to Arong (1966), Moirang to Sendra; Thanga to Keibul and road under construction in between Mayang Imphal to Toubool (Bishnupur) has led to the cutting up of the Loktak into many pieces. And those separated pieces of Loktak wetland areas have been converted into agricultural land. Similarly Loushipat, Poirou and Waithou pats are also being cut into pieces, spoiling the sustainability of the wetland ecosystem. Now, we the present generation should think for a better wetland ecosystem to be inherited by our future generation. It is high time to think together for a better and sound environment for the restoration of wetland eco-system in the region.

Wise Use of Our Wetlands

Wetlands provide tremendous economic benefits. Some examples are water supply (quality and quantity), fisheries (over two-thirds of the worlds fish harvested is linked to the health of the wetlands), agriculture - through the maintenance of water tables, timber production, energy resources - such as peat, plants matter, and electricity, wildlife resources, transport recreation and tourism.

Wetlands also have special attributes as part of the social and cultural heritage of humanity. They are related to religious and cosmological belief systems, and constitute a source of aesthetic inspiration, provide sanctuaries for wildlife and form the basis of important local traditions. This is especially true for the indigenous communities in Manipur.

These functions, values and attributes, can only be maintained if the ecological processes taking place in wetlands are allowed to continue. Unfortunately, and in spite of important progress made in the recent decades, wetlands continue to be among the world's most abused and threatened ecosystems. This is because of on-going artificial drainage, conversion, and pollution by chemicals and ill-conceived over-exploitation of their resources.

"At the threshold of a new millennium, the Ramsar Convention is making a critical review of its achievements to date and setting a clear course for the future, stressing the need to integrate the conservation of wetland biodiversity with sustainable development (considered synonymous with the conservation's concept of 'wise-use') and the health and well being of people every where. In all regions of the world, human population are suffering social, economic and environmental hardships resulting from the destruction and mismanagement of the natural resources base in particular wetlands".

13

Loktak Lake

Loktak lake is, one time, the biggest fresh water natural lake in the Eastern India. It spreads over an area of about 278 km^2 during the normal monsoon. However, during the high flood season it reaches the maximum expansion of about 490 km^2. The probable genesis of the Loktak lake seems to be caused by a tectonic movement resulting in the upliftment of the Manipur river bed, a few metre after the confluence of the Chakpi river near Serou. Some are of the opinion that the present valley; once part of a very much bigger lake, gradually silted up leaving the present Loktak and its sister lakes in the south and south - eastern portion of the Manipur valley. Geomorphologists and Geographers believe that the Loktak lake is part and parcel of the Manipur river drainage system. Eminent paleo - botanist like Dr. Vishnu Mitra expressed the view that the Loktak lake was formed during the middle of the last Glacial Era, i.e. approximately 25,000 years ago. Directly or indirectly the Loktak and its sister lakes support the existence of various bio - diversities of the region. It acts as catalytic agents of the whole range of natural eco - system in the region and also allowed to transmit various wave lengths of socio - politico - economic and cultural activities in the region. Over and above the Loktak lake is an important source of Bio - energetics for the entire state of Manipur and consequently a large section of rural population in the vicinity depends on it for their livelihood (H. L .Sarkar) Considering with all these vital roles played by the Loktak lake, a study of the entire ecosystem with Loktak as the epicentre will not reflect the excessive rate of deforestation and soil erosion, the existing pattern of land use and population pressure on land in the state but also reflecting the role of projects implementing in the Lake region. Now, it is high time to consider all the basic angles and focus attention on the gravity of the environmental crises that the people of Manipur face. The peculiar physical features of the state give rise to queer environmental attributes. This has been further aggravated by the past and on - going environmental abuse. These attributes are constantly undergoing a change from being highly favourable to highly adverse. It is said that upto the census year 1951 the state maintained an optimal population in relation to the available land & natural resources of both hills and plain of Manipur. During this time, it is presumed that the rate of deforestation and jhum cycle were most probably in the normal trend which allowed rejuvenating the forest area. But after 1961 the state's population growth suddenly jumped up causing tremendous pressure on land and forest of the state. As we know well that almost all the valley bound river or the catchment area of the Manipur river are excessively exposed to the continuous actions of deforestation. As a result a huge amount of surface run off with high rate of silt and debris from the hill slopes were brought down during the monsoon. All these debris brought down from the catchment area were partly deposited to the river beds and to the lakes, correspondingly rivers and lakes bottom of the valley become shallower and shallower. This alarming situation is apparent from the serious doubts about the water holding capacity of the Lake as we see currently. The maximum water spread area of the Loktak is not likely to be 490 km^2 as was found 4/5 decades ago. The total water spread area at the flood time may now be only about 286 km^2 (panwar). During sixties, large areas along the lake fringe had been physically delinking from the main Loktak Body. As for example construction of surface road in between Bengul and Arong (1966), Moirang to Sendra Hillock; Thanga to Keibul and road under construction in between Mayang Imphal to Bishnupur are cutting the Loktak into many pieces. And those separated pieces of Loktak lake have been converted to agricultural land. These areas were encroached by the vested interest group of people. This situation has further been aggravated by the Takmu fishery firm by converting a portion of the lake into small pond with earth filling. By these combine process of cutting lake into pieces and encroached upon the cut pieces converted all the shallow lake fringe area into agricultural field. Further along the newly constructed road inside the Loktak lake has been continuously occupied by human settlement.

During the late sixties and early seventies people of the state were fully hypnotised by the Loktak Mulipurpose Project with three boons i.e. (i) 60,000 hectares of land will be reclaimed in the vicinity of the Loktak Lake to be distributed to the landless agrarian population; (ii) 40,000 hectares of land will be irrigated for intensive cropping and (iii) After the commission of the project per unit consumption of energy will be supplied to consumer at the rate below 10 paise per unit. The state was happy with all these expected facilities and day dreaming of growth of pollution free small and medium scale industries with the power available from the Loktak Multipurpose Project. In the totality the standard of living in the state will be raised up. On the contrary, with the commission of Loktak Multipurpose Project and its main organ Ithai Barrage, a huge area of arable land all around the lake has been put under the inundated water for all the time. Further, the impounded water of the Ithai Barrage created a chain reaction over the Pumlen and Lamjaokhong area. By these effects, one of the oldest settlement area of Chairel has been forced to shift towards the nearby hill slopes etc.

Because of the construction of Ithai Barrage, the normal natural behaviour of the Manipur river drainage system were terribly affected, resulting to all river beds of the upper Ithai rapidly shallower and shallower. Natural flow of the Ungamel has been completely altered to the opposite direction. The Khordak cut allows only inflow of debris water towards the Loktak. This process is further accelerated by the shortening of Jhum cycle (2/3 years cycle) and extensive deforestation in the main catchment areas of the Manipur river drainage system. As a result more than 4.4 lakh c.u.m of silt sediment is being brought into the Loktak every year (LDA). Over and above, the increasing rate of shortening Jhum cycle and extension of deforestation rate in the catchment area of the Manipur river drainage system accelerated the rate of surface runoff and soil wash from the hill slopes towards the valley carrying flash flood. Consequently the flood waters first flowing into the Loktak lake and deposit much of their debris load in the lake. It is for this reason that the lake is silting up at an alarming rate from the periphery inward. This is most menacing threat to the Loktak lake and Loktak Eco - System. Now, knowing all these changing circumstances of the Loktak Lake and the Environmental condition of Manipur, many sensitive individuals, NGOs come up with the action of protection of our Environment and to Save Loktak Lake. Taking the chance of the people's opinion some agencies like LDA started the task for dredging the Loktak Lake. Based on the cooked data of the WAPCOS, the LDA already procured three imported dredging machines expending crores of rupees. But by this action of dredging and its dredged earth materials piled up and rising the shallow lake peripheries of lake makes a higher reclaimed land. By this process peoples are ceaselessly reclaiming such peripheral land under swamp and phumdis for cultivation cum settlement. Generally the technique used for such reclamation is to use a combination of flood prevention bunds and the flood drainage channels. In fact the trends of these activities of reclamation are almost on a free for all basis, the advantage of which is being taken, ironically, more by people of means than by the real landless poor in dire need : a large chunk of land are taken possession of and the reclamation process done by dredging. By this time, if some one goes to the side where currently dredging work is going on, it will be vividly visible that a huge area were rising near the Karong of Yangoi Achouba and Yangoi Macha. And simultaneously many people encroached the piled up area for permanent settlement with poultry, kitchen gardening etc. This is happening just at the centre of the Loktak Lake. Same thing the dredger is working at Moirang to Sendra Road side for widening the road by approximately 30 feet. This space will also be encroached very soon.

Thus the action of dredging the Loktak lake is on allowing the lake in converting a pool and also at the same time inviting many more people to encroach upon the Lake area. Thus the problems of Keibul Lamjao, Wild Life Sanctuary are also affected in many angles. Issues are complicated in the areas affected by the construction of Ithai Barrage. Thus many more people's organizations like All Manipur Ithai Barrage People's Organization, Centre for Environment Protection Group, other groups against the use of chemical pesticides for fishing and also protection of Sangai etc. sprang up. The effort of these organizations needs people's support to reach the ultimate goal for a systematic and sound environment in the state. Therefore the people who are born and brought up in this part of the globe and willing to die in this soil need to take care for saving the Loktak lake. The people of this region only can save the environment of this region. No other people from outside the area will not save Loktak. Hence, Loktak Lake needs people's care urgently.

14

Waterfowl Census in Manipur

Introduction

The systematic monitoring of the wetlands, such as lakes, estuaries, and other inland and coastal water bodies, whether perennial or temporary, has become an extremely important activity contributing to a wholesome understanding of our national and international biomes. Various national and inter-governmental specialized agencies worldwide have taken up numerous activities for this purpose. Our understanding of the close and complex inter-relationship between the healthy status of our water resources and the well-being of all forms of life is growing. However, the translation of this growing body of knowledge into practice and behavioral changes within communities lags behind. In this area of local change and action, the recognition of our waters as habitat and the enumeration of waterfowl through regular study can be said to be a vital and sustainable sphere of environmental conservation action, strengthening a community participatory Education and Public Awareness (EPA) approach. Such an approach would also contribute to efforts to constantly update information on wetlands of international importance, of which the Loktak is one.

This third round of the annual mid-winter waterfowl census of the Loktak Lake, conducted from 10-11 January 1998, hopefully contributes towards fulfilling the continuing need to monitor the multifarious facets of this rare and unique wetlands in Manipur. This Census has many aspects, the most important one being the vital need to build a holistic understanding of this biosphere, which is an integral part of the larger ecosystem of the Manipur Valley and its surrounding catchment areas. Another important aspect of the census is to strengthen the eco-friendliness and awareness of environmental conservation among the communities that live and depend on the Loktak wetlands through a constructive and inter-active engagement process.

In fact, the process of local environmentally rehabilitative action led to the initiation of a pilot waterfowl census for the first time in January 1996, the report of which was submitted to the Ministry of Environment, Government of Manipur. The pilot census, organized by MASS and supported by the Rhino Foundation for Nature in NE India, met with unexpected success. For the first time, it was possible to numerically estimate the strength of the host of winter visiting birds to the Loktak. It also made all of us who are concerned about our Loktak, painfully aware of the state of deterioration in the environment of the wetland which has resulted in the progressive diminution of the numbers and species of avian visitors that have traditionally wintered here. The pilot census covered about a third of the lake area and enumerated only about two thousand birds, which was a shocking revelation of the situation. Many of the birds that the elders of the communities that live on the fringe of the Loktak named and told us about could not be seen at all.

A second census followed in early February 1997. Due to a number of logistical and timing of the census reasons, a smaller area could only be covered and the enumeration reflected the technical and other difficulties of such operations. Only about five hundred birds were counted, consisting of about nine species.

After the disappointing results of the second round census, the third round census this year has been most encouraging. Almost five thousand birds were enumerated in the operation that covered about thirty-five per cent of the lake. The success of this census may be attributed to several factors, which need to be more thoroughly examined in a systematic manner. Moreover, we received very positive feedback on the local actions of the many committed individuals and organizations.

We have made efforts to improve the quality of the census, one of which was to undertake a pre-census exercise to orient the enumerators according to existing international guidelines. Many constraints that were identified during the first two rounds of 1996 and 1997, however, could still not be overcome. These can be said to negatively contribute to the quality of the operation.

Location

The third round mid-winter waterfowl census was conducted in the Loktak Lake, a designated Ramsar Convention (Convention of Wetlands of International Importance) wetland habitat site since 1990. Other Ramsar sites in India include Chilka Lake, Keoladeo National Park, Wular Lake, Harike Lake and Sambhar Lake. The Loktak is the only wetland area that survives today in a state that still largely reflects the character of the once-extent vast wetlands system of the Imphal Valley. It has been included in the Montreax Record, "a record of Ramsar sites where changes in ecological character have occurred, are occurring or are likely to occur" maintained by the Ramsar Convention Bureau in consultation with the Contracting Party (India) concerned. It is located in Bishnupur District in the southern part of the Manipur Valley and its present coordinates are 93°46' - 93°55' and 24°25'-24°42' N, though it is subject to quite extensive seasonal variations.

The census covered about 35% of this area. The areas covered were Ngakrapat, Biraharipat, Melheibi, Leihaokom, Chinghoupat Thingom, Moirang Hiram Kachin, Phiraan Makhong, Tauthabi Lamyai and Tauthabi Mathak, and Yangoi Maril. The areas of the Loktak have regularly been selected for census becuase it is reportedly the area of the highest number of sightings in casual observation. In this census, an additional reason for selection of this site was for the purpose of comparison with the resulting earlier censuses.

Observations

The census was preceded by a preliminary visit by a team on 14 December 1997 from the Manipur Association for Science and Society (MASS) to finalize logistical and other arrangements. The census was conducted over two days, 10-11 January 1998. A fifteen-member census team, consisted of a coordinating team from MASS and the Global Science Club, Khoijuman, Bishnupur; a video team from EFTV, Imphal; an independent correspndent and columnist; a correspondent from the North-east Sun; and members of the All Burma Students' League who were keen observers and photographers, took part this year.

Two teams who carried out the counting took up the enumeration operation and identification on four traditional dugout canoes. The entire team spent the night of the 10th of January afloat on two phumdi khangpok (huts) on the Loktak Lake. Excellent weather conditions, quite unusual at the time of the year, greatly assisted the census work.

On the first day of the census, the team secured all logistical arrangements and spent the rest of the day briefly reconnoitering, on canoes, the census area selected. Early on the second day, the actual day of enumeration, the team had a pre-census briefing before setting out in two teams. At the end of the census, the teams gathered together for de-briefing and tallying of the results.

It was found that the total number of waterfowl on the site had greatly increased since the first and second rounds of censuses in 1996 and 1997. This could be partially attributed to the enhanced awaraeness of the phumdi khangpok dwellers (fisher-folk) and their pro-active action to prevent poaching, trapping and killing of the birds during the past years, which could have made the lake a much more hospitable and safer wintering quarters. The teams also observed that many birds continued to feed and play around phum dwelling communities. Many feathers were found to be afloat nearby this phum khangpok, as evidence of the fact that the people who live on the Loktak to eke out a life from fishing do not harm these birds. As reported by these people, the trappers are from remote fringe villages, and sometimes even from far away Imphal !

One enthusiastic bird-lover and conservationist, Mr. Oinam Birathoi, who is a phumdi khangpok dweller and who hails from Thanga Oinam Leikai, recounted his and his colleagues' efforts to discourage the practice of trapping and killing the birds. He showed the team a plastic bag containing small poisoned fish that was used to trap the birds, and this was handed to the team. Synthetic Organo-phosphorus insecticides and other pesticides, such as Dimecron and Thionel, are used by the trappers he claimed. The team has submitted the poisoned fish to the Chief Conservator of Forests and Wildlife, Manipur, who has sent the sample to the Forensic Science Laboratory, Pangei, Manipur for examination. The team was also informed that this practice of trapping and killing the avian winter-visitors is still prevalent, particularly around the Keibul Lamjao National Park and Moirang.

It was extremely distressing to learn, too, that the meat of an unidentified deer was reported to be available on sale in the vicinity of the census area early last year. There seems to be a need for further strengthening the efforts by the Wildlife Department of the government of Manipur.

Another interesting observation was that the practice of trapping flying foxes, known locally as Shekpi (Indian Flying Fox, Pteropus giganteus), was widely prevalent in the village of Khoijuman. This is a practice of very recent origin, as the census team found no such practice during the first pilot census of 1996. The villagers use tall bamboo poles, strapped to tall trees, to which are tied thin nylon nets usually used for fishing. Large numbers have been thus trapped so that, now, no flying foxes visit this area anymore. It was reported that some doctors told a local woman who had been diagnosed to have tuberculosis that her disease had progressed to an extent where conventional medical treatment was useless. She consulted a local health practitioner who cured her by feeding her the meat of the flying fox. After this incident, villagers have taken to trapping and selling the animal for its claimed miraculous medicinal properties.

Our general observation include the increasing scarcity of a healthy growth of waterweeds, known as charang. The growth of water hyacinth has also dramatically been reduced in the census area. These and other natural flora seem to provide and attract abundant food for the birds, as very few birds were found resting or feeding in areas devoid of these plants. These changes may be related to the changing hydrological regime and chemical content of the Loktak today. Dredging and manual clearing of too much vegetation for various purposes may need to be carefully evaluated for their overall impact.

The number of phumdi khangpok dwellers, we were informed, has been increasing over the years. The exact numbers today or the increase over the recent years could not be ascertained. We found that the dwellers freely use the waters for defecation as well as for drinking purposes. The distance between the makeshift toilets and their water collection points for drinking and other purposes, because of the extremely limited area of the phumdi upon which they live, was found to be an average of five meters. There could be extensive fatal contamination of the drinking water. Some practical and simple, yet affordable, solution to this perceived problem needs to be developed for this fishing community.

Constraints

A number of shortcomings were highlighted in the report of the Loktak Lake Mid-Winter Waterfowl Census, 15-18 January 1996. These were :

1. Lack of equipment like telescope, binoculars, field camera with telescopic lens, etc.
2. Inadequate pre-field classroom exercises for the fieldworkers about identification of the birds.
3. Inadequate time.
4. Inadequate (or no) funds for the purpose.

Very few of these initially identified constraints have been overcome by the time of this third round of census. Lack of appropriate equipment greatly hampered the operation this time, too. Prefield briefing was carried out; but this could have been better planned, combining classroom exercises with pre-census field trials. The question of inadequate time is connected to the methodology of such a census. The ideal time frame would be to conduct a rapid census that covered the entire area of the Loktak Lake simultaneously, utilizing an adequate number of competent teams. Such a methodology would reduce duplication, and help arrive at a more realistic figure of the actual numbers of wintering birds in the entire Loktak wetland habitat. Reliable enumerators with experience in identifying different waterfowls are still few, limiting simultaneous team's enumeration.

Lack of appropriate funds is finally the main stumbling block in overcoming many of these technical constraints. However, the consistent efforts initiated will definitely contribute substantially towards raising awareness of the significance of the lake and its importance as a habitat for all forms of life.

Conclusion

The fact that nearly five thousand birds have been enumerated during this census clearly reflects major achievements. Firstly, the overall status of the lake as a wetland habitat and winter quarters for migratory waterfowl seems to be improving - the census figure shows a percentage growth of at least 150% since 1996. This is an encouraging rate of increase over the two intervening seasons, considering that the census areas has remained more or less constant.

Secondly, awareness of the need to protect the sanctity of the Loktak as a safe wintering spot for many migratory waterfowl has also risen. From anecdotal evidence, local communities of phum dwellers as well as the villages of the northern fringe have taken up strong action. However, the practice of trapping and killing these visiting birds still continues, and this seems to be particularly prevalent in the southern reaches of the wetlands. Strangely, the area in the vicinity of the Keibul Lamjao National Park, where conservation action sponsored by the state has been most concentrated, seems to be a vulnerable zone as far as the killing of wildlife is concerned. This raises some deep concerns about the strategy that has been adopted over the years, though a more through investigation confirming this anecdotal evidence is called for. The greatly increased number found in the northern reaches of the lake - Biraharipat, Taothabi Lamyai and Moirang Hiram Kachin, has been found in tandem with increased conservation-oriented local community action. There is clearly an indication that consistent and practical measures can be adopted and fostered through a "partnerships for change" approach with the local communities in all reaches of the lake. As the continuing sangai meat trade and recent onset practice of trapping flying foxes demonstrates, awareness and actual changes in community behaviour, attitude and practice can be very complicated.

The role of phum dwellers in actively protecting and rehabilitating the Loktak as a habitat cannot be over-emphasized. This is because, these fisher-folk are the true natural custodians of the Loktak. Their efforts to form a waterfowl protection group grows out of a genuine love for these birds, and this must be recognized, encouraged and supported.

It is hoped that the value of this annual community participatory mid-winter waterfowl census, as illustrative of a locally viable and wise-use approach of wetlands, will be acknowledged widely within the communities of Manipur, institutions and the government. Efforts need to be made to explore, recognize and develop innovative Education and Public

Awareness (EPA) strategies and interventions at the local and national levels. Persisting technical constraints which can be largely overcome with adequate support from appropriate institutional sources, such as the Ministry of Environment, Government of India, the IWRB and the Asian Wetlands Bureau, would also contribute tremendously towards increasing the quality of the census as well as the EPA approach that is developing.

References

1. Provisional Report - LOKTAK LAKE MID-WINTER WATERFOWL CENSUS : 15-16 January 1996, Manipur Association for Science And Society/Rhino Foundation for Nature in North-East India, Gauhati.

2. Loktak Lake Mid-Winter Waterfowl Census - 1997, Manipur Association for Science and Society.

3. Convention on Wetlands : Strategic Pln 1997-2002.

4. List of Wetlands of International Importance - updated 21 March 1997.

5. Ali, S. (1964), THE BOOK OF INDIAN BIRDS, Bombay Natural History Society : 9th Edition (Revised) 1972, Bombay; 162.

6. Ali, S. (1996), THE BOOK OF INDIAN BIRDS, Bombay Natural History Society; 12th Revised and enlarged edition, Bombay : 354.

7. Ali, S. & Ripley, S.D. (1995), A PICTORAL GUIDE TO THE BIRDS OF THE INDIAN SUBCONTINENT, Bombay Natural History Society, Bombay : 165.

8. Sonobe, K. : A FIELD GUIDE TO THE WATERBIRDS OF ASIA; Wild Bird Society of Japan, Tokyo, 1993 : 224.